The Origins of Life and the Universe

The Origins of Life and the Universe

Paul F. Lurquin

COLUMBIA UNIVERSITY PRESS / NEW YORK

COLUMBIA UNIVERSITY PRESS

Publishers since 1893

New York   Chichester, West Sussex

© 2003 Columbia University Press

All rights reserved

Library of Congress Cataloging-in-Publication Data

Lurquin, Paul F.

The origins of life and the universe / Paul F. Lurquin.

p. cm.

Includes bibliographical references and index.

ISBN 0-231-12654-9 (cl. : alk. paper) — ISBN 0-231-12655-7 (pbk. : alk. paper)

1. Life—Origin. 2. Cosmology.   I. Title.

QH325 .L87 2003

576.8′3—dc21

2002035166

Columbia University Press books are printed

on permanent and durable acid-free paper.

Printed in the United States of America

Designed by Lisa Hamm

c 10 9 8 7 6 5 4 3 2 1

p 10 9 8 7 6 5 4 3 2 1

To Hubert, Ilya, Jacques, Jean, Lucia, René, and Victor, without whom I would know very little

Contents

Preface

Step 1: "To be is to do"—Socrates
Step 2: "To do is to be"—Jean-Paul Sartre
Step 3: "Do be do be do"—Frank Sinatra

—COMPILED BY KURT VONNEGUT JR.

My fascination with the origins of life dates back to my undergraduate years as a chemistry major at the University of Brussels, Belgium, in the early 1960s. The city of Brussels had at that time a large bookstore that specialized in the sale of books from the Soviet Union. Possibly for propaganda purposes, the books sold there were very inexpensive (always an attractive feature for college students) and included titles from famous scientists like the biologist A. Oparin and the astronomer I. Shklovskii. I bought for a mere pittance, in French translation, Oparin's "L'origine et l'évolution de la vie" (The origin and evolution of life) and Shklovskii's "Univers, vie, raison" (Universe, life, reason) ($1.10, hardbound). Reading these books, I became convinced of two things: first, questions relating to the origin of life and the cosmos itself were not silly, and second, physics and chemistry, possibly more than biology, held clues to the answers to these questions.

Like all chemistry students, I took the mandatory courses in statistical thermodynamics and quantum mechanics (the latter from the 1977 Nobel laureate for chemistry, Ilya Prigogine) as well as an elective course in nuclear physics. The subject matter of these courses was a lot more sophisticated than what we learned in more biologically oriented courses, although the emerging science of molecular genetics looked quite elegant. The recently published operon model of François Jacob and Jacques Monod fell in that category. Later, this interest in molecular biology led me to obtain a Ph.D. in biochemistry for work on transfer RNAs.

Problems relating to the origin of life and that of the universe were never mentioned by anyone during my university studies. The 1953 Miller–Urey "gas-zapping" experiment and its variants still stood alone in the experimental quest for origins. In the completely different area of cosmology, the Big Bang theory was in its infancy and the discovery of the cosmic background radiation was just barely made. There seemed to be no connection between cosmology and biology. I progressively forgot about my interest in the origin of life and the universe, and concentrated on my work, the genetic engineering of plant systems. Much of this work was accomplished during my tenure as genetics professor at Washington State University.

More than 30 years after my interest in the origins of life first emerged, the field had changed dramatically. The "standard model" of physics had painted an exciting picture of the origin of matter, while theoretical and experimental work in biology had provided new hypotheses for the origin of complex organic molecules, protometabolism, and the nature of the first genetic material. Phylogeny, the study of genetic relatedness between organisms and their common ancestry, had made gigantic advances thanks to DNA sequencing techniques. Everything seemed to be jelling. I then decided to offer a new undergraduate course entitled Origins of Life, which would force me to brush up on the subject and rekindle my old excitement.

Today the search for the origins of life is on strong footing. The National Aeronautics and Space Administration (NASA) and the National Science Foundation (NSF) both fund this type of research in the biological and physical sciences. Even though the number of researchers is not very large, this field has made a strong impression on the public because it deals with fundamental issues of human existence and nature. Space exploration, more than ever, takes seriously the search for extraterrestrial life. Many await with impatience the launching of probes to Europa, one of the four Galilean satellites of Jupiter, to explore its putative ocean for the presence of life. Likewise, Mars missions are being designed to check for the presence of past and present life, especially since new findings have allowed for the existence of subterranean liquid water there.

Here on Earth, molecular biologists are making much progress in the study of the catalytic properties of RNA and its analogs. Chemists have discovered ways to synthesize amino acids at high pressure and temperature, mimicking conditions present in oceanic hydrothermal vents. Exobiologists have discovered complex organic molecules in space and cosmologists now know something about the beginning and the evolution of the universe.

Student interest in my Origins of Life suggests that the general public, as well as undergraduate students, would be interested in a synopsis and synthesis of all these discoveries—hence this book. Since older professors, nearing the end of their career, have the liberty (at least in my view) to leave the drudgery of forced hyperspecialization, I decided to organize my course notes into this book and to have fun writing it. It was indeed fun, and I am glad to share this exhilaration with interested readers.

I do not pretend to be an expert on all the matters covered in this book; this would be impossible, given the diversity and vastness of knowledge that humans have gathered in the domains of physics, chemistry, and biology. However, I believe that eclecticism is not necessarily a vice, especially for a scientist desirous of communicating ideas about science to the interested public. My main objective while writing this book was to show that the logic of science can be used to make deep sense of the world, from the creation of the universe to the creation of life and its diversification.

Kurt Vonnegut Jr.'s quick compilation of nearly 2500 years of philosophy, which I quoted at the beginning of this preface, humorously suggests that we may not yet be close to a grand synthetic theory of our origins and a persuasive explanation of the meaning of life (step 3). What scientists have discovered, however, goes well beyond Frank Sinatra's famous utterings, as I hope this book demonstrates.

I thank the Honors College and the College of Sciences of Washington State University for partial financial support of this project, and my honors students for their unabashed and cogent critique of the first draft of this book, as well as for their useful suggestions. I also thank my wife Linda Stone for carefully editing the text and making pertinent remarks on areas requiring clarification, and for help with graphics. Albert and Jo Gysels, Michael Sinclair, and three anonymous reviewers went over drafts of the manuscript with a fine-tooth comb. Their help has been invaluable. I also thank Robin Smith, senior executive editor for the sciences at Columbia University Press, for encouragement, support, and insights that definitely improved this book. Marjorie Wexler, my copyeditor for a second time, is deeply thanked for an exceptionally good and efficient job. Finally, Professors Leon Radziemski (Washington State University) and Neil Comins (University of Maine) checked and improved the physics and cosmology chapters. As always, all remaining errors are mine.

The Origins of Life and the Universe

Introduction

I feel that what distinguishes the natural scientist from laymen is
that we scientists have the most elaborate critical apparatus for
testing ideas: we need not persist in error if we are determined not to
do so.

—PETER MEDAWAR, *The Threat and the Glory* (1990)

For us humans, the phenomenon of life exists for certain
in only one spot in the universe: here on planet Earth. In
spite of much speculation on the possible existence of
life-forms on Mars and in other star systems, we still have no evidence that extraterrestrial life exists or has existed. We do not know for sure how life began
on our planet, since no one was present when life first appeared, more than 3.5
billion years ago. How then, can we even hope to tackle the problem of the origins of life? How can we reconstruct the past? We do know with a great degree
of certainty that living organisms, from simple bacteria to human beings, all
function in very similar ways. We also know from the fossil evidence that the
oldest living cells were microorganisms (the descendants of which are still extant today) and that more complicated organisms appeared later. These facts
serve as premises and justification for a scientific search for origins. Since the
mechanisms of life are basically identical across all species, it is legitimate to assume, as a working model, that all living creatures share a common ancestor, the
progenitor of all life as we know it. The central question then is, what was this
progenitor and how did it appear?

It has been said that perhaps the ultimate goal of science is to understand the origins of life on Earth, because such an understanding would encompass our own human origins, from the very beginning to the present. We would finally know *what* we are. To begin with, life-forms are material objects, and all matter is made of atoms. Thus a discussion of the origins of life should include a discussion of the origin of matter and the atoms that compose it. Also, since we are sophisticated aggregates of very old atoms, some of which (hydrogen) were made in the newborn universe, and some of which (carbon, oxygen, nitrogen, phosphorus, and others) were made in stars, a full portrait of the origins of life must also include a tracing of the origin of the universe and that of the stars within it.

Perhaps surprisingly, scientists know a lot more about the origin of the universe and that of matter than about the origins of life. The Big Bang model for the creation of the universe, which entails the spontaneous formation of atoms that compose all matter, is now widely accepted by the scientific community. Such a "simple" model does not yet exist for the creation of life, hence my use of the plural in the word "origins" to indicate that several models have been proposed. The existence of several different models for the origins of life does not necessarily reflect contradiction. Rather, this means that scientists are presently not sure which model is more valid. It is therefore important to understand what a scientific model really is, how it is built, and on what basis it can be accepted or rejected. For this, we must go back to the very foundations of science and its inner workings. Without a thorough grasp of the scientific method, the reader would be unlikely to give more credence to scientific models than, say, to models invoking magic or derived from folk tales.

THE MEANING OF SCIENCE

The somewhat arrogant-sounding quote from a book by Peter Medawar (1960 Nobel laureate for medicine) was used at the opening of this introduction to set the stage: science is essentially self-correcting. Science is also governed by a set of rules. These rules, however, are not written in a form that would make them comparable to, say, the rules of chess or those of football. By and large, one could say that science proceeds in a stepwise fashion that includes discovery, hypothesis formulation, and hypothesis verification. What do these terms mean?

Much has been written about the process of scientific discovery and its meaning, often in the biographies of prominent scientists such as Einstein and Newton, for example. However, I cannot recall from such works a single instance of a practical recipe that would guarantee significant contributions to any area of science. It is very likely impossible to encapsulate what scientific discovery and creativity really are. At the root of discovery sits the ability to connect thoughts and observations in a logical fashion, as well as the questioning of what seems to be established knowledge.

It is a characteristic of science that, over time, the scope and depth of its explanations expand dramatically. Einstein's model of the universe (a twentieth century theory) is quite different from Newton's (a seventeenth century theory), *but* the Einsteinian universe is a generalization of the Newtonian universe and in fact encompasses it. Thus Einstein did not prove Newton wrong, he reinterpreted (among other things) the law of gravitation and made it a property of space itself. Newton had never considered this possibility, making his model of gravitation more restricted than Einstein's. Nevertheless, the Newtonian universe is directly deducible from the Einsteinian universe using the appropriate mathematics. This example shows that a deep rethinking of a satisfactory theory (Newton's) can lead to a totally different view of the universe (Einstein's) in which space is no longer just a void without physical properties. We will see later how this new view led physicists to start thinking about an expanding universe and to abandon the idea that the universe is static and immutable.

Here is an example from the life sciences: for about four decades, genes were thought to be made of protein, not deoxyribonucleic acid (DNA). In 1944, it was demonstrated that the protein hypothesis was wrong and that genes are, as we know, made of DNA. Scientists now think that the first genes, which came into existence billions of years ago, were made of ribonucleic acids (RNA), close cousins of DNA but equipped with very different chemical properties.

The revisiting of the two models just described (gravitation, and the nature of genetic material) occurred because scientists formulated new hypotheses to explain known or new phenomena. For a variety of reasons, these scientists were not satisfied with existing explanations and created their own, new approaches; that is, they built new hypotheses aimed at explaining natural phenomena in a better way. Often, these new hypotheses were counterintuitive, and, as a rule of thumb, common sense intuition is not a very good guide for progress in the sciences. We will see several examples of this in this book.

Hypothesis building is the central process of science. Basically, a hypothesis is an idea that germinates in the brain of a person interested in explaining a natural phenomenon. However complicated the process of hypothesis formulation may be, it must be followed by this simple rule: verification or rejection of the new hypothesis must proceed through experimental observations. This is sometimes called "falsification," meaning that a hypothesis is scientific only if experiments can be designed to support or refute it. A hypothesis, or concept, or idea, is not scientific if it cannot be tested experimentally. For example, the proposition that Earth is flat is a scientific hypothesis because it is falsifiable. Of course, we all know Earth is not flat, as clearly demonstrated by hundreds of artificial satellites in orbit around it. On the other hand, saying that a particular written text must be true because it was inspired by a divinity is not scientific because this statement escapes verification. In scientific language, one would call such a statement a postulate, which, by definition, is an undemonstrable proposition. On the contrary, we will see that the concept of early RNA genes is a hypothesis amenable to experimental work.

Then there is the concept of *theory*. In common parlance, a theory is often construed as an unverified idea, an explanation thrown out of nowhere and used to support (or reject) whatever happens to be the subject of conversation. For example, many a conspiracy theory has been advanced to explain the assassination of President Kennedy or the accidental death of Princess Diana. Proponents of these "theories" invariably refer to "suppressed" evidence kept away from the public. Well, since by definition suppressed evidence is not available for consideration, it cannot be referred to in support of such a theory. A better term for a conspiracy theory in a situation like this would be *hypothesis*.

In scientific terminology, a theory is an ensemble of verified hypotheses, or, as the American Heritage Dictionary puts it, "a theory is organized knowledge applicable in a wide variety of circumstances devised to *analyze* and *predict* the nature of a specified set of phenomena" (italics mine). In other words, a theory is an established model possessing a great deal of credibility. For most scientists, the Big Bang is no longer a hypothesis—it is now a respected theory. On the other hand, John Kennedy's assassination and Princess Diana's death cannot be attributed to conspiracies since there is no available evidence to support these claims. However, people should feel free to pursue the idea of a conspiracy hypothesis if they so desire. And by the way, those who decry Darwin's evolution by natural selection by calling it "just a theory" do this idea a great favor by upgrading it from the rank of hypothesis! Likewise, the Big Bang theory is solidly anchored in the physical sciences and is supported by many experimental observations.

Finally, some theories are considered so complete and encompassing that they have earned the designation *law of nature*. Scientists are reluctant to use the term *law* too freely, because this word is so definitive and therefore not well adapted to the fluid nature of scientific knowledge. There are very few laws of nature; just to name one, the law of gravitation (in both the Newtonian and Einsteinian interpretations) has never been questioned. It is comforting to realize that laws of nature can be constructed, but it is also exciting and perhaps a little scary that new observations could always question the validity of a law of nature and bring about a new view of the physical world.

In conclusion, science relies on the ability of the human brain to organize knowledge, based on experimentation, into a coherent, logical, and self-consistent continuum able to analyze and predict the nature of observable phenomena. Science is thus self-critical and open ended. It is not made by fiat and it does not rely on dogma. Predictably, some theories and even some laws will be proven incomplete, which will keep this human adventure alive and well. Science can never offer absolute certainty; that is not its ultimate goal.

Science has sometimes been declared to be just another system of values (skewed toward Western worldviews) and has been accused of providing humans with yet another mumbo-jumbo explanation of the world. And indeed, for the unaware reader, a phrase like "the cosmos may have originated from quantum fluctuations of the vacuum, in accordance with the uncertainty principle" may sound like "in the beginning a great big turtle decided to bear the weight of the world, and this turtle summoned all other turtles to bear its weight plus the weight of the world, so it was turtles all the way down." The big difference again is falsification: if the proponents of the turtle hypothesis want to prove their point, they must provide evidence for the turtles supporting the cosmos. This has not happened, and other cases of dogma-based systems have failed similarly. On the other hand, quantum fluctuations of the vacuum have been measured and the uncertainty principle (see chapter 1) is a law of nature.

Unfortunately, science can be very hard, and it often challenges our intellectual abilities. It forces us to use our brains and shies away from easy, dogmatic explanations. Maybe this is why many find it dry, painful, and unrewarding. It also just does not provide us with ironclad certainty in any area of everyday life. But again, this is not the aim of science. Science can be coldly impersonal, but its logical consistency (all humans share logical reasoning) can be pushed so far as to constitute a system that can explain the world in an objective and reasonable way. The mumbo jumbo of science is based on our brain's ability to comprehend the world around us. This takes attention and studying rather than blind faith.

Let me now introduce some concepts that are at the center of this book. As I said earlier, all life is composed of matter. This is of course a truism, and we think we know intuitively what living matter versus nonliving matter is. As someone once said, to know the difference between a live horse and a dead horse, all you have to do is give it a good kick. But is it really that simple? I think not.

WHAT IS LIFE?

Erwin Schrödinger, one of the inventors of quantum mechanics and the 1933 Nobel laureate for physics, published in 1945 a little book with exactly that title: "What Is Life?" This was a courageous endeavor on the part of Schrödinger, the physicist, because, as he wrote, "A scientist . . . is usually expected not to write on any topic of which he is not a master." He went on with an apology in which he stated that he was running the risk of making a fool of himself by writing a book outside his professional area. Far from making himself look ridiculous, Schrödinger asked and answered the following deep question: can life be accounted for by physics and chemistry? The answer (in 1944, when the book was written) was no, but Schrödinger went on to say that there is no reason to doubt that some day, physics and chemistry *will* be able to account for the events taking place in living cells.[1]

Over half a century later, this day is very near, if not already here. The nature of the gene is now well understood, and many metabolic cascades necessary to maintain an organism alive have been unraveled. Geneticists have even been able to roughly estimate the minimum number of genes that are necessary for a microorganism to be considered a life-form. This number is about three hundred (human beings and other vertebrates, as well as higher plants, contain tens of thousands of genes). Life has not yet been created in the test tube, but many reactions occurring in living cells can be duplicated in the laboratory. Furthermore, the old adage *omne vivum e vivo* (all that which is alive originates from life) is no longer tenable because there were no life-forms (cells) when the universe was created. The origin of cells must therefore be found in inanimate, nonliving matter. This does not mean that the nineteenth century scientist Louis Pasteur, who proved once and for all that spontaneous generation of life did not happen, must turn in his grave. In his days, the universe was mostly seen as having existed for all eternity, and questions about the origin of life were posed in very different terms.

But, then, what *is* life? In a nutshell, as the American astrophysicist Eric Chaisson defines life, it is "an open, coherent, spacetime structure maintained far from thermodynamic equilibrium by a flow of energy through it—a carbon-based system operating in a water-based medium, with higher forms metabolizing oxygen." Life can further be characterized, in a more detailed but perhaps easier to understand way, by the following properties (in a list inspired by definitions provided by Belgian biochemist Christian de Duve):

- The cell is the unit of life. All life on Earth is based on cells—that is, envelopes that contain within their boundaries all the necessary machinery to effect growth and division. Many organisms are unicellular (bacteria, yeast, and even some marine algae that can grow to several feet long), while others are multicellular (sponges, mosquitoes, palm trees, and dogs, to name a few).
- Cells must extract energy from their surroundings to carry out their life functions and power their metabolism. The ultimate source of energy is the Sun, whose photons of visible light are converted into chemical energy through plant photosynthesis mediated by chlorophyll. Among other things, photosynthesis generates the energy-rich molecule adenosine triphosphate (ATP) that is used as a helper in many metabolic reactions. Organisms unable to perform photosynthesis (and hence which cannot use the Sun's energy directly) eat plants or other organisms that themselves depend on plant food.
- Cells must manufacture their constituents from available food. This process is called metabolism and consists of large numbers of chemical reactions taking place inside the cells.
- The metabolic machinery of cells must receive proper instructions to function in a coordinated, precise manner. These instructions are stored in genes, themselves made of DNA.
- Cells must regulate their metabolic activities to fit their environment. Regulatory mechanisms are very complicated and only partially understood. Some regulation takes place at the level of genes and some at the level of the metabolic reactions. Key elements here are interactions between regulatory proteins and DNA, and interactions between the various proteins that make metabolism possible.

- Cells multiply—that is, they divide and produce more of themselves. This is possible thanks to the mechanism of DNA replication and the complicated reactions that drive cell division.
- Finally, cells must adapt to a changing environment (variations in the nature and amount of food, temperature, salinity, and so forth). They do this by mutating their genes at random, the best adapted mutations being selected by the environmental forces at work. Mutants that are not adapted to a new environment are destined to disappear. Mutations are thus the raw material of evolution by natural selection. The randomness of this process is reminiscent of the concept of *chance* in the universe, as first stated by Democritus in the fourth century B.C.E. (see the next section).

Clearly, life is an extremely complex, integrated phenomenon. It is very unlikely that cellular life appeared all at once, given the astronomical odds against such an occurrence. And indeed, the position of science, as we will see here, is that it did not.[2]

We know today that living organisms are composed of chemical elements found everywhere in the universe, even in intergalactic space. Two questions should thus be posed: (a) what is the origin of elements? and (b) how is it possible that these elements organized themselves into structures that obey the rules that define life? The scientific answers to these two questions constitute the core of this book. But before getting into our scientific views of creation, let us examine some early interpretations of the origins of matter and life.

SOME ORIGINS MODELS BEFORE SCIENCE

Science as we know it is a seventeenth century European invention. This is not to say that earlier Europeans and earlier non-Europeans did not engage in scientific activities. The ancient Greeks, from Thales of Miletus in the sixth century B.C.E. to Galen in the second century B.C.E., accumulated a wealth of knowledge in astronomy (just to name two cases, the size of the earth calculated by Eratosthenes and the Earth-to-Moon distance figured out by Hipparchus), mathematics (Pythagoras, Euclid), and biology (Aristotle). Similarly, astronomy flourished in the Mayan empire in Mesoamerica and in the Mauryan empire in India well before the seventeenth century. Nonetheless, hypothesis-driven science, practically indistinguishable from our modern science, appeared in the 1600s and is associated with the names of Galileo Galilei and Isaac Newton, both physicists.

The problems of the origins of life were addressed scientifically only later—for example, by Charles Darwin in the nineteenth century (he envisioned a little pond where heat, light, and electricity would act on dissolved salts and drive chemical reactions leading to the formation of proteins) and Alexander Oparin of the former Soviet Union in the twentieth century. Oparin was the first to hypothesize, in 1924, that organic compounds necessary for the appearance of life could have been synthesized in the earth's atmosphere. Darwin's and Oparin's hypotheses are still alive and well today (see chapter 4). As for the origin of matter, it was not seriously debated before the 1940s, after physicists had acquired a reasonably good grasp of the nature of the atom.

Of course, questions on the origin of life and the cosmos were formulated well before the development of modern science. Practically every society has built stories to explain its own origins, and these stories almost invariably involve religious ideas. Interestingly, these early myths did not consider the possibility that life could have arisen in a very simple form and acquired complexity over long periods of time. Rather, complex organisms are created by a deity or deities and do not experience evolutionary changes. Also interestingly, the three examples examined below postulate that the universe (or Earth) originated from a shapeless, unstructured substrate.

The Hindu tradition (ca. 2000 B.C.E.), in one of its forms, holds that at the beginning of one of the creation cycles, after a dissolution into nothing, there was no heaven, no Earth, and no space. A vast dark ocean washed upon the shores of nothingness (the void). On this ocean floated a giant cobra in whose coils slept the Lord Vishnu. From the void then came Om, the Sacred Syllable (figure I.1), growing in strength and energy. This awakened Vishnu and the dawn of creation broke. From Vishnu's navel appeared a lotus flower, and sitting in its middle was Brahma, the god of creation. Vishnu then ordered Brahma to create the world. Vishnu and the cobra vanished and Brahma used parts of the lotus flower to create the heavens, the earth, and the skies. Finally, Brahma populated the bare earth with living creatures.

The Judaic tradition (ca. 1250 B.C.E.) describes the well-known 6-day-long creation in the book of Genesis. There again, "the earth was formless and empty" and dark; there was no sky, and land and water were not separated. YHWH (THE LORD) brought order to this situation by first creating light, then the sky, and then by separating land from water. Somehow, God created

FIGURE I.I The Hindu symbol, Om.

light before He created the stars, including our own star, the Sun, on the fourth day. Life first appeared on the fifth day in the form of marine creatures and birds, while land animals, plants, and humans were created on the sixth day.

And finally, Greek mythology (before 900 B.C.E.) assumed that in the beginning there was Chaos (empty space) from which Gaia (the earth), Hades (the underworld), and Eros (love) emerged. Gaia begot Uranus (heaven) and Pontus (the ocean), with whom she mated and produced, among others, the Titans (the old gods who ruled before the Olympian gods) and various sea creatures. Gaia is thus the source of all life. Then, the later descendants of Gaia, with Zeus (son of the Titans Rhea and Cronus) at their head, turned against the Titans and defeated them. Zeus emerged as the leader of the Olympian gods and went on to create humans. Displeased by them, he destroyed humanity with a great flood but allowed one man and one woman to survive on a boat. Earth was subsequently repopulated, as in the story told in Genesis.

All three stories concur that the world (in the form of either a void or a primitive Earth) is initially shapeless and later self-organizes to some extent (Greek mythology) or gains structure through the actions of a deity (Genesis, Hindu tradition). In some narrow sense, this idea of a world (or universe) emerging from nothingness (or close to nothingness) is not very far from the modern scientific view of the creation of the universe through a Big Bang. The analogy should not be drawn too far, however, and certainly, modern science and myth part ways where the origin of life is concerned. Neither Brahma, YHWH, nor Gaia plays any role in science.

It is fascinating to note that a series of Greek philosophers, from about 600 B.C.E. to 300 B.C.E., dismissed mythological explanations for the origin of the universe and the origin of life. They basically invented materialism, an approach that tried (and still tries) to find a unified concept of matter from which the whole universe is built. For example, Anaximander of Miletus hypothesized that the cosmos originated from a primordial substance that he called *apeiron* (meaning indefinite). This model is reminiscent of another Hindu interpretation, in which everything originated from *prakrit*, a kind of primitive matter, containing the essence of all things to come. Some other ancient Greek attempts to explain the origin of life had strange premises (such as Empedocles' idea that life-forms first originated from haphazard combinations of preformed organs), but the view of one of these philosopher-scientists, Democritus, still resonates today. According to him, the universe consists of atoms and void (again) and, moreover, everything existing in the universe is the fruit of chance and necessity. Presumably, Democritus meant that the universe never was a preordained thing (it thus came about by chance) and that its contents appeared as an inescapable follow-up to its creation (necessity). Many scientists think today that this is indeed the case. Democritus may well be the greatest visionary of all time.[3]

The scientific method is a distant relative of the type of thinking that the ancient Greeks invented. Like them, modern scientists take a materialistic view of nature and do not rely on magical, mystical, mythological, and theistic principles. This is not to say that all scientists are virulent atheists. Indeed, many have been and are religious. Simply, as the great French mathematician Laplace once told Napoléon Bonaparte, "Sire, God is a hypothesis I do not need." And indeed, science and religion should not be seen as antagonistic; they just do not need each other, as they ask questions and give answers within very different modes of "knowing." Hence, my intentions are not polemical. Rather, I simply want to strictly adhere to the principles of scientific discovery and interpretation of our world in my description of life and the universe. The excellent book by Isis Fry, *The Emergence of Life on Earth: A Historical and Scientific Overview*, contains an extensive discussion of the philosophical and theological questions raised by origin-of-life research. I highly recommend Fry's work to readers once they are acquainted with the basic principles described in *Life and the Universe*.

This book is subdivided into six chapters. Chapter 1 examines the concepts of modern physics and cosmology that provide the foundations for our understanding of the universe as a whole. Chapter 2 describes the Big Bang model of the creation of the universe and matter as well as the creation of stars and planets. Chapter 3 explains what we know about life as it exists today from the viewpoint of molecular genetics. Chapter 4 tackles the problems of the origin of complex molecules that made life possible billions of years ago and examines how increasing chemical complexity led to the threshold of life. Chapter 5 presents hypotheses concerning the appearance of the first bacterial cells and their evolution into more complex eukaryotic cells. Finally, chapter 6 discusses the possibility that life did not originate on planet Earth but first appeared on other solar planets, and perhaps in other star systems. The possibility of present and past life in our own star system, outside of Earth, will also be discussed, as well as the ultimate fate of life in the universe. Ultimately, I hope to show in this book that evolutionary thinking, taken in its broadest sense, is driving the science of the origins. The universe and the phenomenon of life are not static; both change and evolve, and there truly was a beginning.

CHAPTER 1

Foundations of the Universe

The more the universe seems comprehensible, the more
it also seems pointless.
—STEVEN WEINBERG, *The First Three Minutes* (1993)

The universe was not pregnant with life nor the biosphere with man.
—JACQUES MONOD, *Chance and Necessity* (1971)

These two great scientists, Weinberg the physicist, Nobel Prize 1979, and Monod the biologist, Nobel Prize 1965, seem to speak in concert: there is no imperative for the universe to exist, nor is there one for the existence of life, including that of humans. Weinberg goes on to say, "But if there is no solace in the fruits of our research, there is at least some consolation in the research itself," and Monod, writing about the transcendence of ideas and knowledge over ignorance, announces, "[Man's] destiny is nowhere spelled out, nor is his duty. The kingdom above or the darkness below: it is for him to choose."

These two men lay bare the stupefying yet exhilarating recognition that there is no design in the universe. At the same time, there is a great sense of freedom and responsibility that this thrusts upon us. There *is* a human imperative, but it does not originate outside of us; we have created this imperative ourselves. We must now complete our own destiny: our self-imposed search for the origin of the universe and that of life. These are the two deepest questions that science can ask and perhaps answer. Let us now start our quest, from the very beginning, and study the foundation upon which our understanding of the universe is built.

For thousands of years humans have gazed at the night sky, the Sun, the planets, the surface of Earth, the plants and animals, and themselves. They have tried to make sense of what they saw and constructed explanations to justify the existence of the natural world. Most of these explanations have not passed the test of time. Today a set of interlocking scientific theories exists that provide tentative answers to the origin of the universe and that of matter. These theories result from the melding of relativity and quantum physics into cosmology. But to understand the current model for the creation of the universe, one must first know what the universe contains and understand the physical theories that made this model, the Big Bang, possible.

WHAT IS IN THE UNIVERSE?

As Carl Sagan, the famous American astronomer and science popularizer once suggested, only the word *billions* can give us any idea of what the universe is all about. The cosmos is very large. Some wealthy people may own a few billion dollars, but even that number, as high as it may seem, is nothing in comparison to the number of stars in the known universe. In mathematical notation, 1 billion is 10^9, 1 followed by nine zeros. The universe contains 10^{11} (100 billion) galaxies, groups of stars, and each galaxy contains on average 10^{11} stars, for a grand total of 10^{22} stars. Therefore the estimated number of stars in the observable universe is 10^{13} times bigger than the number of one-dollar bills in a billionaire's wallet. This means that if each dollar bill were equal to one single star, the cosmos would be 10,000 billion times richer than the richest tycoon. This number has to humble us.

The size of the observable universe is 10^{23} kilometers (the distance from Seattle to Miami is about 5×10^3 km), while its age is 12 to 15 billion years. In our corner of our galaxy, the Milky Way, interstellar distances are huge as well. Our nearest neighbor star, Proxima Centauri, is 5×10^{13} km away. Since it is impractical to use numbers with such large exponents, astronomers use *light-years* to express distances in the universe. In those units, Proxima Centauri is 4 light-years away from us. In other words, we now see this star as it was 4 years ago because it took that long for its light to reach Earth. To get a better idea of what this distance means, our star, the Sun, is only 8 light-minutes away from Earth. The large neighboring galaxy Andromeda is 2 million light-years away. It is so large, about as large as the Milky Way, that it is visible with the naked eye in the constellation of the same name, between Triangulum and Cassiopeia. By peering at the cosmos with our most powerful telescopes, we see the most distant galaxies as they were some 10 billion years ago, since it took about 10 billion years for their light to reach Earth. This means that we will observe them as they are today in 10 billion years from now, if anyone is left to make the observation.

As we consider our solar system, numbers become less impressive. Only nine planets orbit the Sun and humans have sent spacecraft to eight of them (Pluto has yet to be visited), with actual soft landings on two (Mars and Venus). We now believe that planets are probably common in the universe. In recent years, planets have been discovered in several dozens of nearby star systems. These planets have not yet been observed directly; rather, their presence has been inferred by measuring the wobble that their gravitational pull exerts on their star or by measuring the drop in their star's brightness as they transit in front of it (literally partially eclipsing it). As for planets that harbor life, we know of only a single one.

Astronomers and cosmologists have calculated an age for the universe, about 12 to 15 billion years. This date of course implies that the universe had a beginning, now universally known as the Big Bang. Thus the universe has not existed for an eternity. Before exploring how the universe got started, however, it would be good to have a feel for what 12 to 15 billion years represent on a scale that we can understand. Carl Sagan cleverly organized this time span into what he called the cosmic calendar. In this calendar the age of the universe is compressed into one single year, starting January 1 and ending December 31 at midnight. Here is an excerpt of this calendar:

January 1	Big Bang
May 1	Formation of our galaxy, the Milky Way
September 9	Formation of the solar system
September 14	Formation of Earth
September 25	Origin of life on Earth
October 9	Oldest known bacterial fossils were alive
November 12	First complex eukaryotic cells appeared
December 1	Earth's atmosphere is fully oxygenated
December 16	First worms
December 20	First land plants
December 22	First amphibians
December 24	First dinosaurs

December 29	First primates
December 31	First humans (at 10:30 P.M.)
December 31	Domestication of fire (at 11:46 P.M.)
December 31	Invention of the alphabet (at 11:59:51 P.M.)
December 31	Roman Empire (at 11:59:56 P.M.)
December 31	Renaissance in Europe (at 11:59:59 P.M.)
December 31	During the last second of the year, five centuries have elapsed and humans have engaged in space exploration.

It is interesting to note that in the cosmic calendar, it took several months for our Milky Way galaxy to form. We will see in chapter 2 what happened during these "months." Also, planet Earth formed nine and a half months into the year, but life appeared only two and a half weeks later. This is lightning fast on the cosmic scale. We will see in chapters 4 and 5 what may have happened during that time.

As mentioned, the Big Bang model is rooted in theoretical physics and, in particular, in the convergence of relativity and quantum mechanics. Let us now review these theories to understand how scientists have pieced together an intellectually satisfying account of the birth of the universe.

PHYSICS HOLDS ONE OF THE CLUES
TO THE ORIGINS: RELATIVITY

The first quarter of the twentieth century witnessed two great discoveries in physics. In chronological order, they were the realization that mass and energy are equivalent and that matter can be understood both as waves and as particles. The first discovery was made by Albert Einstein (a German, then Swiss, then American citizen) as he was developing the theory of special relativity. The second discovery was made by several scientists who founded quantum mechanics. This new view of matter and energy later allowed cosmologists to put together a comprehensive model for the origin of matter as a direct consequence of the creation of the universe.

Albert Einstein's famous equation $E = mc^2$ (E is energy, m is mass, and c is the speed of light) derives directly from his and others' requestioning of the laws of motion that Galileo and Newton formulated centuries ago. This equation shows that mass is a form of energy. It was Einstein who came up with the concept of mass and energy equivalence in his special relativity theory. The term *relativity*, as used here, deals with the measurement of physical phenomena taking place in frames of reference whose movements are relative to one another, at constant speeds. *Frame of reference* defines any location in space characterized by a set of coordinates (up, down, left, right, front, and back), while a *physical phenomenon* is anything that can be measured. For example, you could throw a ball forward (a measurable physical phenomenon) while sitting in a moving bus (one frame of reference) while your friend, standing still on the sidewalk (a second frame of reference), is watching you.[1] For you the ball simply moves forward at a certain speed, but for your friend it also moves together with the bus. For her the ball moves faster than for you because the speed (or velocity) of the ball and the speed (or velocity) of the bus add up. If you throw the ball toward the back of the moving bus, the ball moves more slowly for your friend than for you because the velocity of the ball must be subtracted from that of the bus. This is what Galileo and Newton had already described and understood.

Einstein, however, took the position that light behaves differently from anything else that moves. For him the speed of light is the *same* in all nonaccelerating (that is, moving at constant velocities) frames of reference. In that view, no matter how fast a frame of reference is moving, the speed of light cannot be exceeded. For example, if you flash a laser pointer toward the front of a moving bus, that light, for your motionless friend on the sidewalk, *will not* move at its own speed (about 300,000 km/sec) *plus* that of the bus. For *both* you and your friend, light will move at exactly same speed, about 300,000 km/sec. The same concept applies to a laser pointing toward the back of the bus. Again, for both you and your friend, light will move at the same speed, not more slowly for her, regardless of the speed of the bus. The constancy of the speed of light has dramatic consequences for physics, as I will next explain.

In classical (Newtonian) mechanics, by definition, a constant force acting on an object for an infinite period of time will impart on the object an infinite velocity. For example, an infinitely large fuel tank feeding a rocket engine burning for an infinite time would make a rocket go infinitely fast. Not so in relativistic mechanics; here an infinite velocity is impossible because nothing can go faster than light. Even at an infinite time, during which acceleration is occurring, the object will simply get closer and closer to the speed of light but will never reach it and certainly not exceed it (figure 1.1). This principle is not just a figment of Einstein's imagination but has been verified experimentally many times. The speed of light in a vacuum is indeed constant, regardless of the ve-

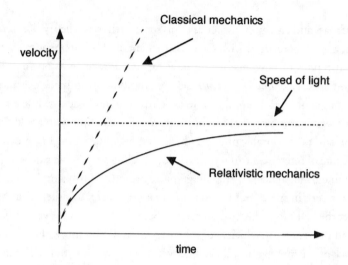

FIGURE 1.1 A graph of velocity versus time, representing acceleration, in Newtonian mechanics and in relativistic mechanics. Relativistic and Newtonian mechanics predict the same values for acceleration at low velocities. At high velocities, relativistic mechanics predicts smaller and smaller values for acceleration, contrary to Newtonian mechanics.

locities of reference frames. (It is, however, not the same in say, air and glass, but the principle of relativity applies under all conditions.)

Next, Einstein realized that time and space are intimately linked and that time is a fourth coordinate, adding to the familiar three dimensions of regular space. Thus in relativity, one speaks of space-time. We know that moving objects move because over time we do not see them exactly in the same position as before. But the speed of light is an intrinsic part of this process because our measurements imply that light moves from the objects to our eyes. Thus space-time becomes a relative quantity (just as positioning in space alone is) that differs in different frames of reference. This complicated reasoning led to the seemingly absurd conclusion of time dilation. According to Einstein's calculations, a clock (the time-measuring device) in a moving frame of reference runs slower than a clock inside a *motionless* frame of reference. And this is exactly what happens! Time dilation has been measured by very accurate atomic clocks installed aboard artificial satellites and fast-moving airplanes. This effect is normally observable only at very high speeds because the equations contain a divisor proportional to the square of the speed of light, a very large quantity. Thus our everyday experiences preclude us from feeling time dilation because most of us do not move at speeds that exceed about 0.3 km/sec (about the speed of a jet), whereas the speed of light is 1 million times faster (see appendix 1).

Now, since space-time is relative (as time measurements show), what about mass? In Newtonian mechanics, mass is a constant that does not depend on the speed of the object possessing the mass. Going back to figure 1.1, we see that in relativistic mechanics, a constant force applied over time to an object will result in smaller and smaller velocity increments experienced by the object. Close to the speed of light, these increments become infinitely small. In Newtonian mechanics a body's rate of change of velocity with time remains *constant*, which means that acceleration remains constant over time. However, in relativistic mechanics the velocity rate of change, the acceleration of a body, *decreases* with time and tends toward zero at infinite time. The body is thus seen as resisting acceleration.

The ability of an object to resist velocity changes is called inertia. Mass, however, is a measure of inertia. For example, a massive rock requires a larger effort to move than a small rock because its inertia, its mass, is greater. Thus in special relativity, the mass of an object increases as its velocity increases, and it tends toward infinity as velocity approaches the speed of light, therefore precluding any object from reaching that speed! The speed of light cannot be reached because at that speed, the mass of the object would be infinite and the force needed to continue to move it also infinite. In special relativity the mass of an object is represented by $m = m_o/\sqrt{(1 - v^2/c^2)}$, where m_o is the rest mass (the mass of the object when it is not moving), v is the object's velocity, and c is the speed of light. As can be seen, the quantity v^2/c^2 appears in the equation, meaning that relativistic effects are negligible at low velocity but become important as v approaches c. Also, the equation shows that if $v = c$, the divisor equals zero and thus m is infinite.

Again, this is not just the result of some complicated mathematical tricks. Subatomic particles such as protons and electrons have been accelerated in "atom smashers" (particle accelerators) so that they move at a rate close to the speed of light. As their velocity increases, it becomes more and more difficult to give them additional velocity because their mass increases, just as the equations predict, increasing the additional force needed to accelerate them. For example, in a synchrotron used to accelerate electrons, the magnetic field needed to deflect electrons moving close to the speed of light is 2000 times greater than that predicted by Newton's theory. This is because at these relativistic speeds, the electrons are 2000 times more massive than when at rest, as predicted by the special theory of relativity (see appendix 1).

What has this got to do with the equivalence between mass and energy? Newton had demonstrated that the energy of a body by virtue of its velocity is $E = \frac{1}{2}mv^2$, where m is the mass and v is its velocity. Remember that Newton regarded mass as a constant. For Einstein, as we have seen, mass is not a constant, since it increases with velocity. Thus the energy E does not simply increase with v^2, it also increases as mass (m) increases with v. Thus mass should be considered a form of energy. But Einstein went further and made the bold assumption that the rest mass of an object (its mass regardless of its velocity, as low as it might be) is *also* a form of energy. Einstein was of course proven right with the detonation of the first atom bomb in 1945, which demonstrated the conversion of a plutonium mass into heat (a form of energy), electromagnetic energy (gamma rays), and nuclear fragments.

More interestingly, Einstein's theory also predicted that energy in the form of radiation could materialize, provided that radiation packed enough energy. That is, special relativity predicted that matter could appear from radiation, which is the reverse of what happens when a nuclear warhead is detonated. This was demonstrated when energetic gamma rays were observed to materialize into electron-antielectron pairs (that is, matter and antimatter). Thus it can also be said that energy is just another form of matter. Thus $E = mc^2$, or $m = E/c^2$. Mathematically this is trivial, but conceptually it is not. In particular, this means that energetic photons (electromagnetic radiation of zero rest mass) of high energy E and moving at the speed of light, as all photons do by definition, can be converted into material mass. In that sense, matter can appear from nonmatter. This mass-energy equivalence has important consequences for the creation of the universe, as we will see in the next chapter.

About 10 years after he formulated his special relativity theory, Einstein published his general theory, in which reference frames were no longer assumed to be in steady motion relative to one another; they could also experience acceleration. Newton had already demonstrated that acceleration is involved in gravitational effects such as those experienced by planets orbiting a star. The general theory of relativity is then a theory of gravity. This theory involves mathematics much more sophisticated than that used for the special theory (which involves only high school math—at least as taught in a good high school). There Einstein demonstrated the equivalence between gravity and the warping of

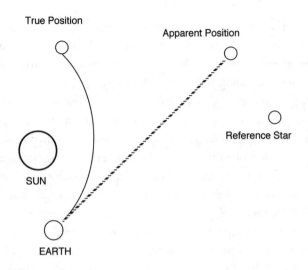

FIGURE 1.2 How starlight is deflected by the gravitational well of the Sun. A photo of a star whose light path grazes the Sun was taken during a total solar eclipse, which allowed this star to be visible during daytime. Another photograph, taken at night without the Sun in the line of sight, showed the shift in the star's position caused by the Sun's deflection of its light. The reference star is far away from the sun and its light is not deflected by the Sun's gravitational well. The angular distance between the true and apparent positions is greatly exaggerated for clarity. The actually measured angular shift is only about 1 second of arc, in full agreement with Einstein's calculations.

space. In other words, gravity was no longer the property of massive bodies per se; these bodies (such as stars) deformed the fabric of space-time and led to the formation of gravity "wells" in which satellites such as planets would "fall" and orbit. Similarly, starlight would be deflected by the gravity wells of other stars (figure 1.2). This conclusion was quickly verified. Massive objects like stars are indeed able to deform the straight path of light emanating from other stars because they warp space. General relativity thus seemed like a good theory to describe the universe, which is of course a vast space full of massive bodies (the stars and galaxies) possessing impressive gravitational fields.

Einstein and Willem de Sitter (a Dutch physicist and a contemporary of Einstein) applied the equations of Einstein's general theory to the universe as a whole, and the result had one important consequence: this universe cannot be stable; it must expand. Newton had already identified the problem of the stability of the universe. In his model, the universe could not be stable either, because the masses of the stars and their gravitational attraction would cause an inevitable collapse, given enough time. In the Einstein-de Sitter universe, on the other hand, expansion was predicted. In 1916, when Einstein's theory was published, there was no evidence whatsoever that the universe was contracting or expanding. Yet his general relativity theory offered the possibility that at one time, the universe's size could have been exactly zero—in other words, a singularity in space-time with infinite energy (remember the mass, that is, matter-energy equivalence). Einstein himself dismissed this possibility, since there was no experimental evidence to prove (or disprove) expansion or contraction at the time he published his work.

Now, Einstein's insight and rigorous mathematical development was proven right when, in 1919, astronomers demonstrated that starlight can indeed be deflected by massive objects such as stars. Was there something wrong with general relativity as it pertained to the whole universe, as opposed to just individual stars or galaxies? No, we know today that Einstein's general relativity is more complete and verified than any other theory (except quantum mechanics, not a competing theory) ever formulated. Einstein's interpretations were not just the ramblings of an ivory tower dweller, they were real. Therefore the whole theory, including an expansion of the universe, must have great validity. The question was, then, is space really expanding?

Careful observation of the movement of galaxies in the cosmos provided the answer and, in a nutshell, this answer is that space is indeed expanding. The expansion of the universe was put on firm footing by several astronomers working in the mid to late 1920s. For this, these astronomers decomposed the light of distant galaxies by using prisms or similar devices. When starlight is decomposed by a prism, we observe the fundamental colors that combine to give white light (in the case of the Sun), and a series of dark lines. These dark lines are called an absorption spectrum; they result from the absorption of specific wavelengths by elements present in the Sun and other stars (figure 1.3A). This is how helium was discovered in the nineteenth century, by noting that some absorption lines in the Sun's spectrum did not correspond to any element known so far on Earth but were distinguishable in sunlight.

Galaxies are composed of many billions of stars, so the absorption spectrum of a galaxy gives an indication of what elements predominate in the stars present in that galaxy. One element whose absorption lines are particularly convenient to observe is calcium. Astronomers made an astonishing discovery: the far-

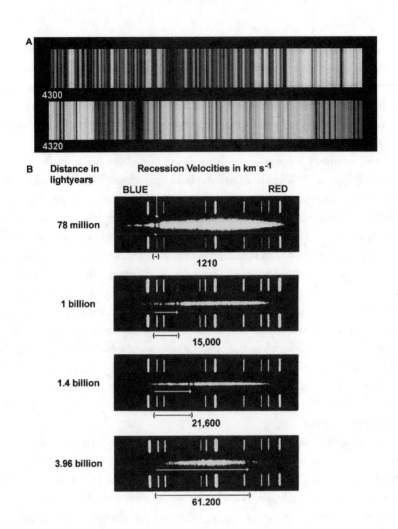

A
4300
4320

B Distance in lightyears

Recession Velocities in km s^{-1}

BLUE   RED

78 million
(-) 1210

1 billion
(———) 15,000

1.4 billion
(———) 21,600

3.96 billion
(—————) 61.200

FIGURE 1.3 Absorption spectra and their use in determining galactic velocities. **A:** A small portion of the absorption spectrum of the Sun (blue region, from 4300 to 4340 angstroms). The many dark lines represent wavelengths absorbed by the many elements present in the outer layer of our star. **B:** The shift of two calcium absorption lines toward the red end of the spectrum as a function of recession velocity. The *white arrows* under the spectra indicate the magnitude of the redshift. The four galaxies represented are, from the top: a Virgo cluster galaxy, Ursa major, Corona borealis, and Hydra. (Adapted from Weinberg, S. 1993. *The First Three Minutes*. New York: Basic Books.)

ther away the galaxy, the more the calcium lines are shifted to red wavelengths. When stationary (as in a laboratory situation), calcium's prominent spectral lines are in the blue region of the spectrum. However, these calcium lines shift to the center of the spectrum, in the yellow-green region, for galaxies not too far away, and to the red region of the spectrum for very distant galaxies (figure 1.3B).

Now, calcium on Earth is the same as calcium elsewhere in the universe. What is happening? Why are all spectral lines, including those of calcium, red-shifted in distant galaxies? This redshift can be understood in terms of what is called the Doppler effect. This effect is familiar to all of us. When an ambulance is using its siren (or bell, or other sound device), the pitch of the siren increases as the ambulance approaches us, then decreases as the ambulance moves away. This is because the sound waves of a moving object increase in frequency as the object is approaching but decrease as the object is receding. The same happens with light waves: when a luminous object approaches us, the frequency of the light it emits is shorter (blue-shifted) than that of the same stationary object (figure 1.4). Conversely, when the object moves away from us, its light is red-shifted. What is more, the amount of blueshifting or redshifting is proportional to the velocity of the object. (However, do not expect to see blue-shifted or red-shifted car lights; cars move much too slowly for the Doppler effect to be noticeable with light.)

Applying the Doppler effect to the shifting of calcium lines (and other elements' lines) in the spectra of galaxies, astronomers were left with only one conclusion: the spectral lines of galaxies are red-shifted because the galaxies are receding from us at high speed, and their wavelength is increased relative to stationary spectra. What is more, they discovered that the more distant a galaxy is, the faster it recedes.[2] For example, the Virgo cluster galaxy, which is 78 million light-years from us, recedes at 1200 km/sec. On the other hand, the Hydra cluster, which is 4 billion light-years away, recedes at 61,000 km/sec. Therefore Einstein's prediction of an expanding universe now had an observational foundation: since galaxies recede from one another, the size of the universe is growing with time; it is expanding. This expansion should not be misunderstood: the universe is not expanding *into* something such as a big void; it is space itself that is expanding. This effect actually "stretches" the wavelength of light, making it redder because red light has a longer wavelength than blue light.

The expansion concept immediately suggests that the universe had a beginning. If, indeed, we ran the clock back and looked at the universe earlier and

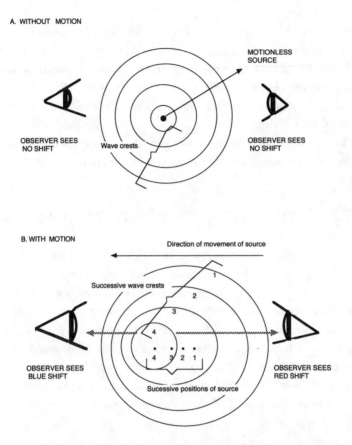

FIGURE 1.4 The Doppler effect. This effect is caused by the compression (or expansion) of the wave crests of light in the direction of motion of a light-emitting object. If the object is moving toward the observer, the waves are compressed, generating a shorter wavelength (a higher frequency), which shifts the original color toward the blue end of the spectrum. If the object is moving away from the observer, the waves are stretched, generating a longer wavelength (a lower frequency) and shifting the color toward the red end of the spectrum. This effect depends on the velocity of the object. **A:** Wave crests from a motionless source are not shifted. **B:** The wave crests of a moving source are compressed in the direction of motion and stretched in the opposite direction. The light of the most distant galaxies is redshifted to such an extent that they are most observable in the infrared portion of the electromagnetic spectrum.

earlier in time, the galaxies would be seen to approach one another, make contact, and eventually perhaps merge into a Big Crunch. As this crunch progressed, we would witness an enormous increase in density, as all the matter in the universe would be concentrated into a smaller and smaller volume. Could the end of this process be the zero-volume singularity mentioned earlier in this chapter? Could all these billions and billions of tons of galactic matter concentrate into a very small volume? Before addressing this question, we must first understand what matter is and how matter and radiation interact.

THE NEXT CLUE: QUANTUM PHYSICS

The macroscopic world in which we live is deterministic. When we see a moving car on the road, we expect to be able to precisely determine its location and speed. This is how the police deliver speeding tickets. Not so in the world of the atom, where indeterminacy is the norm. Also, a solid object such as a brick is obviously composed of material elements, not electromagnetic radiation such as light, which we normally characterize by its wavelength or its color. Again, not so in the atomic and subatomic world, even that making up the brick. We will indeed see later that particles of matter can also be characterized as waves. The branch of physics studying phenomena at the level of the atom and below is called quantum mechanics, the field that has generated quantum indeterminacy and matter waves.

Quantum mechanics was developed in the middle and late 1920s but had its origin at the turn of the twentieth century. Einstein, together with Max Planck (of Germany), Louis de Broglie (of France), and Niels Bohr (of Denmark), all Nobel laureates, founded this new science. However, a second generation of researchers, Max Born and Werner Heisenberg (both Germans), Erwin Schrödinger and Wolfgang Pauli (both Austrians), and Paul Dirac (an Englishman), also all Nobel laureates, gave quantum mechanics its present face, one that some call quantum weirdness. This is because quantum mechanics defies our common sense; it is completely counterintuitive. Nevertheless, quantum mechanics and its derivatives, quantum electrodynamics and quantum chromodynamics, are exquisitely precise in measuring, interpreting, and predicting the behavior of atomic and subatomic particles.

Quantum mechanics and relativity theory are the two pillars of modern physics. These are two theories that simply work extremely well to explain the physical world. Modern cosmology, the study of the universe on a grand scale, uses both relativity (which deals with universe-size objects) and quantum mechanics (which deals with the inframicroscopic world) because, as one could surmise, the properties of matter and radiation must have an effect on the universe, which contains both.

The birth of quantum mechanics occurred in 1899, when Max Planck demonstrated that light energy was not exchanged with matter in a smooth continuum, as had been thought until then. By smooth continuum, I mean a mechanism that does not operate by jerks. For example, when I depress the accelerator pedal of my car, I admit more fuel into the cylinders, where it burns. That chemical energy is turned into kinetic energy at the level of the pistons, which in turn operate the axle and finally the wheels. This chemical energy and the kinetic energy are not produced by leaps and bounds; smoothly depressing the accelerator results in a smooth acceleration.[3]

In Planck's day this was what everybody in the world of physics assumed to be true: light, a form of electromagnetic energy, and matter interacted in a perfectly linear way, not in a stepwise fashion. But there was then the nagging problem of the blackbody radiation. A blackbody in physics is a hollow, material object able to absorb and reemit, as electromagnetic radiation, the thermal energy it has received. Thus a blackbody heated at high enough temperature can turn red hot, then white hot, much like a piece of iron. The blackbody then emits visible light that the human eye can perceive. If heated at a higher temperature, the blackbody starts emitting even shorter-wavelength light—namely, ultraviolet (UV) light.

It turns out that the theoretical equations used in the nineteenth century to model the behavior of a blackbody as a function of temperature were hopelessly flawed. These equations predicted that at high temperature, the blackbody would emit an infinite amount of UV radiation. This was not experimentally observed. In fact, the amount of UV radiation was observed to drop off, not increase to infinity. Clearly, the theory did not match observations, a damning fact for a theory. Planck set out to derive an equation that matched experimental observations, and he succeeded in his efforts. To his own surprise, he discovered that his new equation implied that electromagnetic energy could not be emitted by the blackbody in a continuous, smooth fashion. Rather, his equation predicted that the energy came in discrete, finite packets or bundles, which he called quanta. Therefore energy emission of the blackbody was not unlike the steps of a staircase: there is no such thing as half, a quarter, or an eighth of a step. Like steps of a staircase, quanta came in full units of themselves and this meant that energy could not be emitted on a smooth, continuous scale divided into infinitely small increments. Energy in the form of electromagnetic radiation was received and emitted in jerks—that is, in discrete amounts.

Planck's work was entirely theoretical, but Einstein demonstrated in 1905 that quanta can explain very real experimental observations. To do this, Einstein solved the mystery of the photoelectric effect. Some metals, such as sodium, release electrons when they are irradiated with light. This is the photoelectric effect. The released electrons, when forced to move by an electric field such as that imposed by a battery, produce a current. This principle is at work in elevator doors that reopen as an entering passenger is on the verge of being crushed.

Before Einstein it was known that the intensity of a current (the number of electrons released by the metal surface) was determined by the intensity of the light reaching the metal. However, and this is what puzzled everybody, the wavelength of the light had to be just right. If the photocell responded to blue light, it would not respond to red light, no matter how intense the red light was. Red light simply refused to release electrons from the metal surface and nobody understood why. Enter Einstein. Although it was contrary to what the scientific community had held for many years, Einstein assumed that light, which was evidently a wavelike phenomenon, could also act like particles or packets of energy. And it worked! Einstein's energy packets, now called photons, were Planck's quanta.

This is how, according to Einstein, the photoelectric effect works. Electrons are bound to the atoms composing the surface of the light-sensitive metal. If the photons of light that hit the metal surface carry an energy less than that of the electrons' binding energy, no electrons will be released. If, on the contrary, the photons carry enough energy, electrons will be knocked off the metal surface and freed. Since the energy of a photon is directly proportional to the frequency of the light in question (through $E = hf$, where f is the frequency and h is a constant called the Planck constant), this explains why low-frequency light (red light, for example) cannot extract electrons from the surface, whereas higher-frequency light (such as blue light) can do the job (figure 1.5). Simply put, in this example, the photons of red light do not pack enough energy to free electrons, whereas photons of blue light do. Einstein's quantum interpretation of light was fully verified experimentally in 1914 and he received the 1921 Nobel Prize for physics for this discovery.

Now this quantum theory posed a problem: is light an electromagnetic wave or is it made of particles? An electromagnetic wave is a combination of coupled,

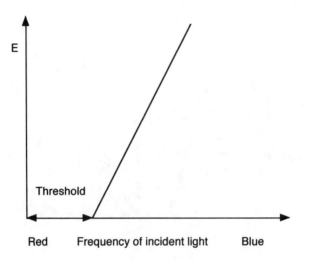

FIGURE 1.5 Graphic representation of the photoelectric effect. The *slanted line* represents the energy of the released electrons as a function of the frequency of the incident light. The threshold represents the frequency below which no electrons are emitted. In this graph, red light is located to the left of the horizontal axis and blue light to the right.

pulsating electric and magnetic fields created by a moving electric charge (figure 1.6). One example of an electromagnetic field is that generated by electrons moving back and forth in a radio antenna. Thus radio and radar waves; infrared, visible, and UV light; X rays; and gamma rays constitute what is called the electromagnetic spectrum, from low frequency and low energy (radio waves) to high frequency and high energy (gamma rays). Newton first proposed that light rays were made of particles, but subsequent experiments suggested a wavelike nature for light. The wave nature of light was put on strong theoretical footing by the physicist James Clerk Maxwell in the second half of the nineteenth century. However, this theory failed to explain the photoelectric effect! The revisiting of the wave/particle duality by Einstein became, as we just saw, a turning point in physics. We know today that light can be seen as packets of electromagnetic energy (particles called photons) *and* as waves. Its corpuscular and wavelike nature can be revealed in different experiments, using different instruments; there is no contradiction between these two aspects of light. The dual nature of light is now fully accepted by all physicists.

But, then, if light is both radiation and particles, what about matter? Could

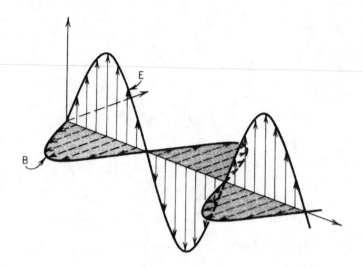

FIGURE 1.6 An electromagnetic wave showing the coupled electric (E) and magnetic (B) fields as they oscillate in their motion through space.

it be that the particles of matter also could be seen as some kind of wavelike phenomenon? The answer of quantum mechanics is a loud yes. Thus light has a dual nature, but so does matter. This is how this other duality, that of matter, was demonstrated. By the 1910s, scientists had shown that matter is composed of atoms containing a central body, the positively charged nucleus carrying most of the mass of the atom, and electrons, negatively charged particles, somehow in orbit around the nucleus. The diameter of the nucleus is very much smaller than the diameter of the atom. To get an idea of the proportions involved, the nucleus has been compared to the head of a pin sitting in the middle of a football field, the borders of the atoms (its electrons) being located at the sidelines.

It had been known since the nineteenth century that a gas could be "excited" by an energy discharge (such as an electrical spark), in which case it would emit light. Interestingly, the spectrum of the emitted light consists of discrete bright lines on a dark background (figure 1.7). It was also known that gases, in an "unexcited" state, absorbed light to generate absorption spectra that were, to simplify, the negative image of emission spectra. Absorption spectra, as we saw earlier, are used to measure the velocities of galaxies. Thus gases (matter)

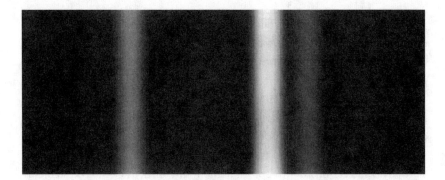

FIGURE 1.7 The emission spectrum of mercury. When decomposed by a prism, the ghostly white color of excited mercury (as in some streetlights) generates three prominent lines in the blue (left), and green and orange (right) parts of the visible spectrum.

could emit and absorb electromagnetic energy represented by light of a specific frequency.

In 1913, the physicist Niels Bohr proposed that atomic spectra were the result of orbiting electrons releasing energy in the form of photons (quanta of light) when excited in some way, by electrons, photons, or chemical reactions, for example. In this scenario the electrons absorbed the provided energy and reemitted it in the form of photons by dropping back to their original, or at least lower, energy level. Absorption spectra were caused by the reverse phenomenon: incident light of a particular frequency was absorbed by electrons that were thus bumped up to a higher energy level. What was exciting about this interpretation was the notion that electrons in matter could receive energy and release it only in a quantized—that is, discrete—manner. Remember that Einstein had demonstrated quantized energy acquisition by electrons in the photoelectric effect but not quantized energy release (which does not occur in the photoelectric effect anyway). Bohr thus proposed the notion of discrete electronic orbits around the atomic nucleus. Electrons acquiring energy in the form of quanta of light are forced into higher, more energetic orbits. Once there they can fall back to a lower energy state (a lower orbit) and emit a quantum of light. These quanta of light are emitted (or absorbed) at very specific frequencies, explaining the narrow (in frequency or wavelength) lines of emission and absorption spectra. This also meant that electron orbits were themselves quantized: an electron in an atom could not assume any energy level, only some such levels were possible. Electronic transitions between energy levels are known as quantum leaps or quantum jumps, expressions that have joined our everyday language.

Problems with this simple (but revolutionary) model soon arose. While Bohr's theory explained well the spectrum of hydrogen (an element containing a single electron around its nucleus), it did not work in the case of another gas, helium, which has two electrons. Another problem was that, according to classical physics, electrons in orbit around an atomic nucleus cannot be stable. If they are thought of as little bodies carrying a negative electric charge in orbit around the nucleus (much as planets are in orbit around the Sun), the theory of electromagnetic radiation demonstrates that electrons, being electrically charged moving objects, must radiate energy and soon fall into the nucleus. Matter then is not stable! But of course, we know that matter *is* stable. Clearly, the electrons-in-orbit model is incomplete or wrong, or merely a first step.[4] What was left of Bohr's model, however, was the idea of quantum jumps occurring in matter and the correlation between atomic spectra and electronic energy levels.

In 1923 the young scientist Louis de Broglie (a genuine duke) published the results of his Ph.D. dissertation, a series of cogitations inspired by Einstein's idea that mass and energy are two facets of the same coin. Since, thought de Broglie, both mass and light are forms of energy, matter (which has mass) and light should be describable using the same mathematical formalism. He showed that, indeed, electrons could be mathematically described as waves, not just as very small material bodies! This great insight was demonstrated experimentally a few years later. (The principle that electrons are waves is used in electron microscopes.) This again posed a problem: what *are* electrons? Are they material objects (they had better be, because they are one of the constituents of matter) or are they waves? Or else, are they both? These questions are of course reminiscent of the particle-wave duality for light that we examined earlier.

The year 1926 saw the grand synthesis of Planck's, Bohr's, Einstein's, and de Broglie's ideas into one single coherent theory: quantum mechanics. Werner Heisenberg and Erwin Schrödinger demonstrated independently (using very different mathematical tools) that electrons can be treated as three-dimensional matter waves surrounding the atomic nucleus. In this picture, Bohr's orbits are replaced by electronic "clouds" and the electrons cease to be seen as single material points revolving around the nucleus, as planets orbit the Sun. In the new model, an electron is "delocalized, smeared out" and the cloud (also called an orbital) represents the *probability* of finding an electron in the region of space occupied by the cloud (figure 1.8). The simplified Schrödinger equation is

$$-h^2/8\pi^2 m \cdot d^2\Psi/dx^2 + V\Psi = E\Psi,$$

where h is Planck's constant, m is the particle's mass, V is the potential energy that depends on the particle's position, x is the particle's position, E is the total energy of the particle, Ψ is the wave function, and $d^2\Psi/dx^2$ is its second derivative. The wave function Ψ has no direct physical meaning, but generally speaking, wave functions that satisfy the Schrödinger equation allow only certain energy (E) values for the particle. These energy values are thus quantized.

The probabilistic interpretation of matter waves was proposed by the physicist Max Born, who realized, soon after Schrödinger and Heisenberg published their work, that mathematically, Ψ^2 is the probability of finding a particle in a given region of space. Thus Ψ^2 is a probability density. When one thinks about this interpretation for a minute, one realizes that it has baffling consequences. For the first time in physics, it had been suggested that one could no longer pinpoint the position of a material object! Indeed, how does one pinpoint a wave or a probability density?

Not only that, the Schrödinger equation allows particles to overcome energy barriers that would be considered classically impenetrable. Let us consider the following "classical" example. Imagine a billiard ball in a strong, tall box. The ball can be pushed around, even energetically, but in no case do we expect the ball to escape from the box. This is because the walls of the box constitute a solid barrier that the ball cannot overcome. This is not what is happening in the world of the atom, because the Schrödinger equation allows subatomic particles to "tunnel" through energy barriers that would have stopped these particles in a classical world. Figure 1.9 describes the tunnel effect. A subatomic particle is shown as a wave present between two potential energy "walls" of value V_0 (the V as in Schrödinger's equation) that classically would prevent the particle from escaping.

In reality, solutions to the Schrödinger equation show that the particle has a nonzero probability of penetrating the potential barrier and finding itself in the outside world. This is as if the billiard ball in our example were able to cross the walls of the box without breaking them. And again, this is not simply the result of clever mathematical tricks. We now know that radioactive α-decay, the phenomenon in which radioactive substances (such as plutonium) emit helium nu-

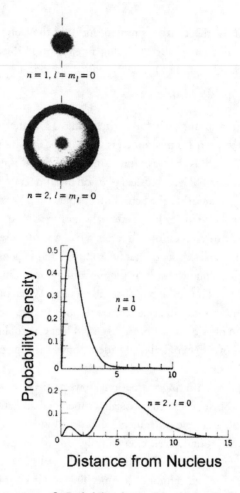

FIGURE 1.8 Probability density "clouds" for the electron of a hydrogen atom. The electron is represented in two different energy (quantum) states, $n = 1$ and $n = 2$. Other quantum numbers l and m describe the angular momentum of the electron. At the lowest energy level ($n = 1$), the electronic orbital forms a fuzzy sphere near the nucleus (the proton). At the higher energy level ($n = 2$), the electron can be found in two different areas around the proton: one very close to it and another, larger one, five times more distant. The curves show sections of the three-dimensional figures. Energy absorption by the electron causes it to move from the lowest energy level to the next, higher one. By emitting a photon, the electron falls back to the lowest energy level. (Adapted from Spielberg, N., and B. D. Anderson. 1987. *Seven Ideas That Shook the Universe.* New York: John Wiley & Sons.)

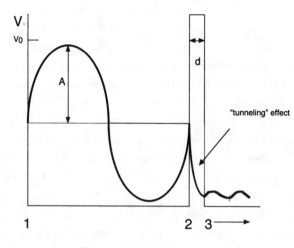

Distance of propagation

FIGURE I.9 The tunnel effect. A subatomic particle is shown as a wave of amplitude A. The potential energy barrier V_0 is such that the particle is confined between "walls" 1 and 2. It can be shown that a wall of thickness d can be penetrated by the particle, which then escapes confinement and travels on from point 3. This is impossible in the "classical" macroscopic world.

clei (two protons associated with two neutrons to form an α particle), is explained by the tunnel effect. Alpha particles are not energetic enough to defeat the potential energy barrier presented by the nuclei of these radioactive elements; they leave these nuclei by tunneling through this barrier.

Quantum weirdness was developed further by Heisenberg who soon enunciated his famous uncertainty principle. According to this principle, it is impossible to determine with infinite accuracy *both* the momentum (the product of the velocity and the mass) and the location (the spatial coordinates) of an atomic or subatomic particle. In other words, if the location of a particle can be precisely determined, its velocity cannot, and vice versa. This has nothing to do with the accuracy of our instruments; Heisenberg demonstrated that indeterminacy is an intrinsic property of nature as described by quantum mechanics. The uncertainty principle is written

$$\Delta p \cdot \Delta q \geq h/2\pi,$$

where p is the momentum (mv), q represents the three coordinates of space (used to position any object in three-dimensional space), h is the Planck constant, and π is 3.14159 . . . , making $h/2\pi$ a constant number. The symbol Δ represents intrinsic imprecision, variation in the measurements. Thus if we measure the particle's position exactly, Δq, the variation in position, will be zero. But if Δq equals zero, then Δp must be equal to or greater than $h/2\pi$ divided by zero. This means that the imprecision in the measurement of the momentum (mass times velocity) is infinite! Heisenberg's uncertainty principle is often misunderstood. The imprecision in the measurements of p and q has nothing to do with imperfect instruments and equipment; Heisenberg showed that his principle is a direct mathematical consequence of quantum mechanics. And again, this principle was demonstrated experimentally.

The Planck constant, h, is a very small number, and hence the uncertainty principle is not observable in the macroscopic world. This is why you should never invoke this principle when a policeman is writing you a speeding ticket.

The uncertainty principle also applies to another pair of variables and can also be written

$$\Delta E \cdot \Delta t \geq h/2\pi,$$

where E is the energy of a particle and t its lifetime in an excited state. Since $E = mc^2$, the uncertainty principle allows for the totally counterintuitive existence of *virtual* particles! Simple algebra shows that particles can materialize from nothing and return to nothing if their lifetime Δt is equal or smaller than $(h/2\pi)/mc^2$. In other words, the uncertainty principle makes possible the creation of particles from nothing at all if they return to nothing within a given Δt proportional to their energy! In fact, following the uncertainty principle, we should think of empty space, a perfect vacuum, as brimming with virtual particles popping in and out of existence. Or, to put it differently, when the uncertainty principle is expressed in terms of energy and time, it shows that particles can materialize from nothing, if only for a very short time (see appendix 2). This phenomenon is known as quantum fluctuations of the vacuum. This also means that a vacuum, empty space, contains energy. And yes, bizarre as it may seem, this so-called zero-point energy does exist. The Dutch physicist Hendrik Casimir calculated in 1948 that zero-point energy should force two metal plates to attract each other when brought sufficiently close together. The *Casimir effect* was tested in the laboratory and verified. Therefore a vacuum contains energy, confirming the "reality" of virtual particles and quantum fluctuations. Again, what may have been seen as sophisticated mathematics without physical basis turned out to be a law of nature. Heisenberg's uncertainty principle rules the subatomic world.

Quantum mechanics, after its successful application to the properties of electrons, was quickly applied with equal success to the cases of the two subnuclear particles, the proton and the neutron. It is also applicable to the short-lived particles detected in cosmic rays and in the products of collisions in particle accelerators. In its 75 years of existence, quantum mechanics has never been contradicted by empirical observations. It is a very good theory of nature. We will see later that aspects of quantum mechanics, in particular the wavelike nature of subatomic particles, as well as the uncertainty principle, have important consequences for the creation of the universe. However, the interpretation of quantum mechanics is not complete, nor is it completely accepted. Physicists continue to argue over its meaning.

FORCES OF NATURE AND ELEMENTARY PARTICLES

Ordinary matter[5] is composed of nuclear particles, neutrons (electrically neutral) and protons (positively charged), that combine to form the atom's nucleus. As we have seen, the nucleus is surrounded by negatively charged electrons that counterbalance the positive charge of the nucleus. Physicists now consider electrons to be truly elementary particles. This means that electrons do not seem to have any kind of internal structure. Protons and neutrons are different from electrons. For decades after their discovery they were thought to be elementary as well. Then theoretical calculations and experiments with large-particle accelerators demonstrated that this is not the case. We now know that protons and neutrons (collectively called *baryons,* meaning "heavy") are composed of three subunits each, called *quarks,* with two "up" quarks and one "down" quark for the proton and one up quark and two down quarks for the neutron. As far as we know, quarks are truly elementary and devoid of internal structure.

Another category of elementary particles known to form ordinary matter, in addition to baryons, is the lepton (meaning light) class, which contains negatively charged electrons and electrically neutral neutrinos. The mass of an electron is about 1/2000th that of a proton or neutron. Neutrinos appear to possess a very small mass, much less than that of an electron. Neutrinos interact very weakly with matter and are produced in a certain type of radioactive decay, called β decay. They exist as free particles in the universe, as they are produced by thermonuclear reactions taking place in stars. The Sun is a profuse source of neutrinos, which, incidentally, are totally harmless for living organisms.

All these elementary particles must interact to produce matter as we know it. They do this through the action of forces that must explain how, for example, positively charged protons present in an atomic nucleus do not repel each other instantly. Three such forces explain the stability of matter, while a fourth force explains the action of gravity. Thus modern physics recognizes four fundamental forces, the strong force, the electromagnetic force, the weak force, and the gravitational force. They are characterized as follows:

- The strong force unites quarks inside baryons. Interestingly, and counterintuitively (for most of us, anyway), the strong force is not that strong at extremely close range, but it gets stronger as the distance between two quarks increases. This explains why free quarks have never been observed. The "left over" of the strong force (sometimes called the nuclear force) unites protons and neutrons inside the nucleus and is stronger at closer range.
- The electromagnetic force governs the interactions between charged particles. It is attractive when two particles have opposite charges (positive and negative) and repulsive when the charges are the same. It is weaker than the nuclear force at close range, explaining why two protons do not repel each other when they are present in the same nucleus but do so once the range of the nuclear force is exceeded. Electromagnetic forces determine the overall structure of matter, as in atoms where electrons surround a positively charged nucleus. The electromagnetic force is 100 to 1000 times weaker than the strong force. Its strength decreases with the square of the distance separating two particles (an example of the well-known inverse square law).
- The weak force is also a type of nuclear force. Among others, it governs β⁻ decay, a radioactive process through which a neutron is converted into a proton plus an electron and an antineutrino. Generally speaking, the weak force determines the interactions between baryons and leptons. The weak force has an extremely short range and is about 1 million times weaker than the strong force.
- The gravitational force is always attractive and is 10^{39} times weaker than the strong force. It governs the interactions between masses and follows the inverse square law, as does the electromagnetic force. We saw earlier that the

gravitational force is in fact complex and results from the warping of space-time by objects possessing mass.

In quantum theory, all forces correspond to fields, regions of space where a force exists, and all four types of field are characterized by a given field particle (with its dual wave/particle characteristic)—hence the existence of a third category of particles to accompany baryons and leptons. Thus the strong force is mediated by gluons, massless particles acting at very short range. The electromagnetic force is mediated by the familiar photons, massless particles whose range is infinite (but varies as the inverse square law). Then, the weak force is mediated by the exchange of Z°, W^+, or W^- field particles between baryons and leptons. The field particles here are very massive and hence very energetic. This makes their lifetime very short, according to the uncertainty principle, and thus makes their range extremely short as well. Finally, the gravitational force is governed by gravitons, massless hypothetical particles, with infinite range, also acting according to the inverse square law.

Where does this lead us in our attempts to understand the origin of matter? Going back to the theory of relativity, we understand that mass (matter) is energy and thus energy is matter. Therefore given enough energy in a system, it is possible to create matter. It is then convenient to express units of mass in units of energy such as electron volts, where 1 electron volt is the energy acquired by an electron submitted to a voltage drop of 1 volt. Using this system, and knowing that the mass of a proton is 1.67×10^{-24} grams, its mass in energy units is 1000 million electron volts, abbreviated as 1 GeV. In comparison, the mass of an electron is only 0.5 million electron volts (MeV). Current particle accelerators and colliders achieve energy levels of about 200 GeV. These machines are thus easily able to create matter from energy; indeed, accelerators can create all sorts of subatomic particles. Therefore the equivalence between matter and energy and the mechanisms involved have been experimentally explored for many years.

Now any system (which can be a small collection of atoms, a human body, a galaxy, or even the whole universe) above a temperature of absolute zero emits electromagnetic radiation proportional to its temperature, very much as a blackbody does, as discussed earlier. In scientific notation, temperature is expressed in degrees Kelvin (also called absolute, where absolute zero is -273° Celsius), abbreviated as K. Therefore a particular type of electromagnetic radiation, characterized by a certain range of wavelengths and energies (radio waves, microwaves, infrared and visible light, UV light, etc.) is associated with the *equivalent temperature* of a system. For example, a temperature of 300 K is close to room temperature and a little below that of a human body. At this temperature, objects, including humans, emit primarily infrared light. This is how night vision goggles work; they capture infrared photons and convert them into visible light. At 10,000 K (10^4 K) or less, matter emits light in the visible range, this visible light being composed of higher-energy photons. At 10^5 K, matter emits UV light, which at 10^8 K becomes X rays, and at 10^9 K, it emits gamma rays that can materialize, as we have seen, into electron-antielectron pairs. This is because the energy carried by gamma-ray photons is high enough to create matter according to $E = mc^2$.

Of course, no solid, liquid, or even gaseous matter can exist above a few thousand K; at these temperatures all matter exists in its fourth state, a plasma, in which atomic nuclei and some electrons are no longer bound. At higher temperatures still, matter ceases to exist and is turned into radiation of very short wavelength and hence very high equivalent temperature. Thus it is possible to establish a relationship between the temperature of a system and the particular state (matter or radiation) this system is in. As far as subatomic particles are concerned, this state directly depends on the mass-energy content of the system. Theoretical calculations and particle accelerator experiments have shown that quarks, the constituents of protons and neutrons, can exist stably only below 10^{14} to 10^{13} K. Protons and neutrons are stable below 10^{12} K, whereas electrons stabilize below 4×10^9 K. Atoms—that is, nuclei surrounded by their electronic clouds—become stable in the range of 3000 to 20,000 K.

In addition, theoretical calculations and experiments have shown that at very high energies (very high temperatures), the four forces of nature are united as one, a unified *superforce* of sorts, because field particles (gluons, photons, the weak force particles, and gravitons) are not differentiated. As temperature goes down, gravity separates out, leaving the strong, electromagnetic, and weak forces together. Then, the strong force separates from the electroweak force, and, next, the weak and electromagnetic forces separate. In that sense, very high energy levels are "pregnant" with the four forces and mass, two prerequisites to form matter.

Thus to build a universe that has a beginning in a very small region of space, one needs extremely high temperatures and energy levels high enough to lead to the production of matter, and these high temperatures must then decrease to induce stable matter particles, and subsequently atoms, to exist. Relativity provides a unification of the concepts of mass and energy (special relativity), and space and gravitation (general relativity). Quantum physics, on the other hand, unifies the concepts of particles and radiation. What is still missing, however, is a unification of general relativity and quantum physics. To achieve this goal, scientists will have to develop a quantum theory of gravity—in particular, a theory that can explain the behavior of space-time, gravity, and energy under the extreme conditions of density that prevailed at the moment of the Big Bang. This has yet to be accomplished. Nevertheless, we shall see in chapter 2 that both theories account very well for the properties of the known universe, although they leave the mysteries of the beginning still unsolved. As we now understand it, the universe must have begun in a tremendous flash of released energy: the Hot Big Bang. This energy then coalesced into material mass, interacting via force field particles, and left a large number of free photons behind. This scenario will be studied in the next chapter.

CONCLUSIONS

Relativity and quantum physics both provide accurate descriptions of the world. The realm of relativity is the very large and the very fast. Both regimes exist in the universe, where galaxies, thousands of light-years in size, recede at tremendous velocities. Relativity is the science of space. Quantum physics deals with the ultramicroscopic, the realm of the atom, the molecule, and their constituents. The cosmos basically contains only two things, radiation and matter, both present in a vast amount of space. By understanding radiation and its interactions with matter (quantum physics) and by understanding that radiation and matter are two forms of energy (relativity), we can build a model explaining the creation of the universe and the matter it contains.

On a final note, it is legitimate to wonder whether quantum physics and relativity have any impact on our daily lives. In the case of relativity, the answer is probably not much (with the exception of nuclear weapons and nuclear reactors, direct spinoffs of $E = mc^2$); its realm seems so far limited to pure science. In the case of quantum physics, the answer is a definite yes. Our entire electronics industry is based on quantum theory. As Leon Lederman, Nobel laureate for physics, once said, "If everything we understand about the atom stopped working, the GNP [gross national product] would go to zero."

CHAPTER 2

Building a Universe

The most incomprehensible thing about the universe is that it is comprehensible.

—ALBERT EINSTEIN

T wentieth-century physics, with its precise description of matter, energy, and radiation, was merged with astronomical observations to provide an account of the birth of the universe consistent with the laws of nature. These efforts resulted in the Big Bang model for the creation of the cosmos. Since we do not (yet) know how to create a universe in the laboratory, some of the Big Bang theory relies on mathematical and computer models. However, the sciences of astrophysics and high-energy particle physics provide much empirical evidence to support the theory in its entirety. In this chapter we will see how the Big Bang model works. We will also see that the formation of stars and planets was a direct consequence of the stabilization of matter, once the temperature of the universe had dropped sufficiently, after a significant period of expansion of the universe.

COSMOLOGY: THE BIG BANG MODEL

We saw in chapter 1 that Einstein's general relativity predicts an unstable universe beginning with a singularity characterized by zero space and infinite energy. Using the equations of general relativity, the Belgian priest and university professor Georges Lemaître suggested in 1927 that the universe could have started from a rapidly expanding "primeval atom," as he called it. By this he meant a very small region of expanding superdense time-space from which all matter and radiation originated. This, however, did not solve the question of infinite energy. Infinities are not welcome in science. In fact, nobody knows what infinite energy is. Therefore the Lemaître-Einstein model, interesting as it may be, cannot be used to make predictions regarding the properties of space-time at time zero, the moment of the Big Bang. Nonetheless, 1927 should be considered the birth year of the Big Bang model, even though nobody called it that when this hypothesis was first put forth. Let us first consider, then, what happened a very short time after time zero and avoid for now the infinite energy problem. We will return to speculations regarding time zero at the end of this section.

An extremely important clue to the birth of the universe was revealed when the temperature of the cosmos was measured. The temperature of intergalactic space is not zero Kelvin (K). Space is actually pervaded by a background radiation of 2.7 K, with an energy distribution like that of a blackbody, and a peak wavelength of 2 mm corresponding to a photon energy of 0.00062 electron volts (eV), placing this radiation in the microwave range (remember that any temperature above 0 K is characterized by a given electromagnetic radiation wavelength). This very weak cosmic microwave background radiation is uniformly distributed in all directions around us and thus fills the universe. Human eyes are, of course, unable to detect electromagnetic radiation below the range of visible light, but our radio antennae *are* able to do so, which is how scientists discovered this universal microwave background. What is its origin? How is it that microwave photons are filling the universe? Is there a correlation between this radiation and the beginning of the universe?

As covered in chapter 1, we know that the universe is expanding (remember galactic redshifts). If we were hypothetically to run the expansion clock backward, we would see that as the universe became smaller and smaller, the microwave background radiation density would become higher and higher, until it reached an enormous value. Also, since space would be shrinking, the microwave radiation would be blue-shifted and gain more and more energy, until it too would reach an enormous level. In a shrinking universe, the microwave background would turn into infrared light, then into visible light, and subsequently into short-wavelength light that human eyes cannot perceive. Since the energy density of the universe would increase drastically, so would its temperature. We can even imagine a point at which the temperature and energy density would be so high that matter particles could no longer exist stably. If we mentally reverse the process, in an expanding universe starting from very high photon density and energy, the radiation becomes weaker and weaker as time goes by. The photon density decreases and so does the energy associated with these photons because of the redshift caused by expansion. Consequently, the temperature of the universe drops, and at a certain point, stable matter can exist.

Thinking along these lines, the cosmic background radiation must represent the "ashes," the "afterglow" of the Big Bang! At one point in time, the young universe must have been a reservoir of enormously intense radiation at extremely high temperature. This raises the question of how old the universe is. Based on the rate of expansion, itself calculated from the velocity of the most distant galaxies observable today, the age of the universe is about 12 to 15 billion years (see appendix 3). Extrapolating back in time, it is then possible to calculate the temperature of the universe when it was extremely small, fractions of a second after the Big Bang. Avoiding infinities "à la Einstein," we can calculate that 10^{-43} second after the Big Bang, the temperature of the universe was 10^{31} K and its radius was 3×10^{-35} meter. This is many orders of magnitude smaller than the radius of a proton, putting the just-born universe squarely in the submicroscopic realm of quantum theory. Also, as matter cannot exist at such a high temperature, it is present as very energetic radiation (or, more precisely, the conversion of radiation into matter does not lead to stable matter because the temperature is too high). Nevertheless, calculations indicate that at that time, the force of gravity was already separated from the three other forces (strong, weak, and electromagnetic), even though stable matter did not yet exist.

It may surprise the reader that gravity existed from the beginning, in the absence of stable matter particles. However, we must remember that energy possesses mass. Therefore at extremely high energies, we find a considerable amount of mass, hence the existence of gravity. Also, at such a high energy density, the curvature of the universe must have been enormous and we must also remember that curvature of space *is* gravity. Gravity was the first force to separate, because as temperature decreased, the mass-energy present in the expanding volume decreased as well.

The rest of the timetable then goes as shown in the following table, where the radius is that of the universe in meters, H represents hydrogen, D is deuterium (the heavy isotope of hydrogen) and He is helium:

TIME	RADIUS (M)	T (K)	EVENT
10^{-35} s	3×10^{-27}	10^{28}	Strong and electroweak forces separate.
10^{-10} s	0.13	10^{15}	Electromagnetic and weak forces separate.
10^{-9} s	0.4	10^{14}	Quarks start to stabilize.
10^{-5} s	3000	10^{12}	Protons and neutrons stabilize.
10^{-3} s	300,000	10^{10}	Deuterium nuclei stabilize.
10 s	3×10^{9}	4×10^{9}	Electrons stabilize.
100 s	3×10^{10}	10^{9}	He nuclei stabilize.
400,000 y	6.6×10^{21}	3000	Electrons are captured by H, D, and He nuclei; radiation and matter are completely decoupled.

As expansion and cooling proceeded, the three other forces (with their associated field particles) separated. Then matter particles with highest mass energy started stabilizing, followed by those with less mass energy. This is because matter is stable when temperature is below its energy equivalent divided by $1.5k$, where k is the Boltzmann constant, a number relating energy and temperature. Thus the stabilization temperature is $T = E/1.5k$. Knowing that $E = mc^2$, and knowing m, the mass of elementary particles, one can calculate their energy equivalent. This value of E is then used to calculate T. Since quarks are more massive than baryons (protons and neutrons), which in turn are more massive than electrons, quarks stabilized first after the Big Bang (when the universe was only 80 cm in diameter, the size of a large beach ball), followed by the baryons and finally the electrons as the universe expanded further and its temperature dropped accordingly. It may come as a surprise that baryons, composed of three quarks, are less massive than quarks themselves. This is because quarks, as they unite to form baryons, release energy—and hence mass—in the process. The same principle is at work in nuclear fusion (see the next section).

It should be noted that experimental results have been gathered up to energies of 200 GeV (corresponding to a temperature of 2×10^{15} K) in particle accelerators. This temperature corresponds to about 10^{-10} second after the Big Bang. Therefore it is fair to say that we understand quite a bit of the physics involved in the transition between radiation and matter in the very young universe. What happened before that time can presently be assessed only by theory, as we will see later.

The timetable of the creation of the universe contains several interesting features. First, it should be realized that protons constitute hydrogen nuclei. H is the simplest element, with one proton surrounded by a single electron. Deuterium is an isotope of hydrogen; it contains a single proton and a single neutron in its nucleus. D has only one electron, just like H, and it has the same chemical properties as its lighter isotope. The second simplest element is helium, composed of two protons and two neutrons (in its most abundant form) surrounded by two electrons. D nuclei stabilized 10^{-3} second after the Big Bang, whereas He nuclei stabilized after 100 seconds. It can also be seen that formation of H, D, and He neutral atoms took place when the universe was 400,000 years old, when the temperature had become low enough for atoms to exist.

It is also at that time that the universe became transparent. Indeed, before that time, starting at 10^{-9} second, the universe was a plasma of independent particles and radiation. In a plasma, light cannot travel very far because of strong interactions with particles. Therefore the superhot young universe may have looked like a dense, luminous fog.

The table thus mentions only two elements created in the Big Bang: hydrogen and helium [a third element, lithium (Li), with three protons and three neutrons was formed in trace amounts]. It turns out that calculations based on quantum mechanics predict that a hot Big Bang should produce these two elements in a precise ratio: 76% H and 24% He (with traces of D and Li). And lo and behold, this is exactly the H/He ratio found in the universe at large today (plus the traces of D and Li)! This observation clearly reinforces the notion that the universe's birth conforms to the Big Bang model. Figure 2.1 provides a graphic representation of the major transitions that have occurred in the universe since the Big Bang.

How does one explain this 76:24 ratio? Protons and neutrons are made in equal amounts in the Big Bang and stabilize at about 10^{12} K. At this temperature, the nuclear force is not strong enough to allow stable proton-neutron interactions. However, as has been demonstrated experimentally, starting at about 10^{10} K, these stable interactions are made possible. But neutrons are slightly more massive than protons (by about 1.3 million eV), meaning that the balance between neutrons and protons will shift toward the latter as the temperature of the universe drops. This is because the conversion of protons into neutrons and vice versa favors protons at lower temperatures. By the time the nuclear force al-

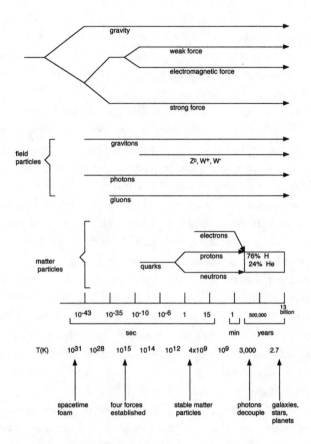

FIGURE 2.1 Diagram showing the major transitions that have taken place in the universe since its birth. The **top portion** shows the separation of what is sometime called supergravity (a single force that existed briefly at the time of the Big Bang) into gravity and a "grand unified force." The grand unified force then separates into the strong force and the electroweak force. Subsequently, the electroweak force separates into the weak force and the electromagnetic force. The field particles that mediate the action of the four forces are also represented. The **bottom portion** of the diagram shows the timetable for the appearance of matter particles and first atoms (neutrinos have been ignored for simplicity). The time and temperature scales are not linear. Photon decoupling from matter indicates the time and temperature at which the universe became transparent to light.

lows proton-neutron binding, there are only twelve neutrons left for eighty-eight protons. The twelve neutrons will unite with twelve protons to make six helium nuclei, each actually originating from the fusion of two deuterons (deuterium nuclei), each containing one proton and one neutron. This step is called nucleosynthesis because atomic nuclei are being synthesized. Then the remaining seventy-six protons will remain alone and will become hydrogen nuclei that later will form the 76 percent hydrogen atoms once these protons each capture an electron. The formation of nuclei heavier than helium (other than the traces of D and Li) cannot proceed because the expansion of the universe, and hence a great decrease in particle density, prevents further interactions between nuclei. This also means that the synthesis of elements of atomic mass higher than 4 (i.e., He) must have taken place elsewhere, and not in the Big Bang. This point will be discussed in the next section.

Taken together, Einstein's unstable universe, the cosmic microwave background radiation, the redshift, and the precise H:He ratio predicted by theory make an extremely solid case for the creation of the universe through a hot Big Bang. Some competing theories, such as a steady-state universe that is expanding but has existed for all eternity, have been presented. However, these theories do not agree with observations (nor can they explain the microwave background radiation) and hence must be rejected.

For a variety of reasons,[1] including the fact that there exist in the universe stars and galaxies, not just the two gases H and He in a uniform and dispersed state, many scientists now accept the idea of an inflationary period that took place between 10^{-33} and 10^{-32} second after the Big Bang. This rapid inflation brought the size of the universe from 3×10^{-27} m to 10 cm. At the end of inflation, the universe had ceased to be a quantum object. By stretching the nascent universe so quickly, it is thought that any subtle preexisting patchiness in the energy distribution would have been maintained because full thermal equilibrium would have been prevented. Once energy converted into matter, this patchiness was kept, and it is recognized today as stars and galaxies. If there was indeed inflation that led to a nonuniform distribution of matter, the cosmic microwave background radiation should also show subtle nonuniformity because this would reflect the presence of heterogeneities from practically the beginning. This was actually demonstrated by microwave detection equipment carried by satellites and high-altitude balloons; the microwave background radiation is very smooth, but it is not completely smooth.

In summary, the standard model for the creation of the universe is very solid. Quantum physics and relativity provide consistent explanations regarding the creation of space-time, matter, forces, and radiation. Inflation seems to explain the properties of the universe as we know it today. However, most people want to know what happened at the Big Bang and what caused it. I was told that one of my colleagues once said in an introductory biology class, "The Big Bang, eh! Who pulled the trigger?" Doubtlessly, he was being facetious. Nevertheless, people and scientists want to know, if not *who* pulled the trigger, at least *what* pulled the trigger. Unfortunately, no simple explanation can be provided yet. What happened at time zero and how it happened are two questions at the ultimate frontiers of science. Theorists are quite active in this area and the following are some propositions they have devised:

First, the universe at the Big Bang may not have had a size of zero. This possibility would violate the uncertainty principle of quantum mechanics. The minimum possible size in the universe's timetable, 1.62×10^{-35} m, is called the Planck length, and it, together with the Planck time, 5.39×10^{-44} second after the Big Bang, are the smallest "knowable" parameters for the nascent universe.[2] Thus time "zero" is not really zero, and the same holds true for zero space. This avoids the problem of infinite energy in a purely Einsteinian universe.

Next, the uncertainty principle allows the universe to have been created from nothing! As we have seen, Heisenberg's principle can be written $\Delta E \cdot \Delta t \geq h/2\pi$. Following this, if the energy content of the universe is zero, the indeterminacy of E is also zero. The consequence of this concept is mindboggling, because if $\Delta E = 0$, then $\Delta t \geq (h/2\pi)/0$, meaning that $\Delta t \geq \infty$, which is an infinity (because any number divided by zero is an infinity)! Thus the universe could exist for an infinite time (not necessarily in its present form, but still composed of matter and radiation) provided it contained zero energy. But *is* the universe's energy content zero? Many cosmologists think that this is the case and that matter-energy (that is, the energy equivalent of all matter in the universe) is exactly balanced by negative gravitational energy. Even though gravitation is always attractive, it can be thought of as a form of negative energy because it binds matter together. Therefore in an expanding universe, where matter is being dispersed, gravitation has a negative sign. The idea of creating a universe from nothing is not completely wild; we have seen that a vacuum is full of virtual particles, as demonstrated by the Casimir effect. In a similar scenario, the universe could have been born from quantum fluctuations of the vacuum, with the universe tunneling (see chapter 1) into existence.[3] In that picture, there is no trigger: nothing but chance caused the Big Bang.

Then there is the question of what happened before the Big Bang. For many scientists this is a philosophical rather than a scientific question. And we have seen that the Planck time limits our ability to go back to true zero. At any rate, it would be extremely difficult, if not impossible, to see what the universe looked like before it reached an age of about 400,000 years. We have a chance to glimpse at what the universe looked like after this time, by devising more and more powerful telescopes that can reach farther and farther into the distant universe at higher and higher redshift. For this, light from these distant objects needs to reach us, and we have seen that the universe was not transparent to light until 400,000 years after its creation. Events before that time are hidden in a quantum fog.[4]

Nevertheless, theorists are speculating on the properties of the universe around the Planck time and have come up with models describing "bubble universes" (or "multiverse"), in which our own universe is but a corner of a total cosmos containing multiple, isolated universes. These bubble universes could be seen as tunneling out of preexisting universes. Others think that time really had no beginning at all. In that view, space-time itself is curved. This allows for space-time to be finite (as imposed by a model involving a beginning for the universe) yet to have no boundary. A familiar analogy is that of the surface of a sphere, which is curved and finite and yet has no boundaries. In that sense, the idea of a singularity at the beginning of space-time disappears. This concept reduces even further the need for a trigger and, together with the uncertainty principle, does away with causality. In that sense, the universe simply *is*.

At any rate, this is not the place for cosmological speculations. We are interested in the origin of matter, because living things are made of matter, and so far, we have seen that the creation of matter is a direct consequence of the Big Bang. However, the Big Bang created only two elements, hydrogen and helium. Hydrogen is very abundant in living matter (representing 63 percent of the human body by number of atoms), while helium is completely absent from molecules that constitute life. In fact, helium is a very unreactive gas and could not have become one of the constituents of living cells. But then where did all the other elements, such as carbon, oxygen, nitrogen, and several metals such as calcium, iron, and magnesium, so prevalent in living cells, come from? We now know their origin: they were made in stars.

THE GENESIS OF STARS AND PLANETS

The mechanism by which stars emit light is well understood. It can be replicated on Earth by detonating nuclear fusion bombs, the most destructive weapons known to humans. Much as fission bombs (the regular A-bombs) are the result of an understanding of the mass-energy relationship provided by special relativity, so fusion bombs (H-bombs) are unfortunately a direct spinoff of our understanding of the way stars shine. Nuclear fusion, as in H-bombs, is the process by which atomic nuclei are forced to fuse under high pressure and high temperature conditions. This process results in the release of tremendous amounts of energy, because atomic nuclei possess less energy than the free baryons composing them. When these baryons (protons and neutrons) are forced together to form new nuclei, the whole contains less energy than the parts. This excess energy is released in H-bombs as electromagnetic radiation and heat (another form of electromagnetic radiation). This is also exactly the way stars function: they fuse atomic nuclei and release fusion energy and light into space.

In the young universe, after hydrogen and helium were synthesized, nuclear fusion would not have been possible without concentrating these gases in parts of the expanding universe. This is where the consequences of inflation may have played a role. The slight variations seen in the distribution of the microwave background radiation possibly correspond to heterogeneities of matter as well, these heterogeneities forming a considerable time before stars and galaxies appeared. It is thought that these heterogeneities gave rise to large clouds of hydrogen and helium synthesized after the Big Bang. These clouds, called protogalaxies, underwent gravitational contraction between about 400,000 and 150 million years after the Big Bang. The general temperature of the universe dropped from 3000 K to 16 K during that period.

However, as the protogalaxies contracted, their temperature increased because the kinetic energy of the infalling gas was converted into heat. This temperature increase is not to be confused with the general temperature of the universe, which eventually dropped to 2.7 K. The temperature increase occurred locally, inside the protogalaxies only. Over time, it must be assumed that the protogalaxies fragmented into smaller contracting clouds, the *protostars*. When the temperature inside the contracting protostars reached 15 million K, the repulsive force between protons (H nuclei stripped by heat of their electrons) was overcome and protons collided and then fused to form helium nuclei. The stars lit up. Now, helium made in stars by the fusion of two hydrogen nuclei contains two protons but also two neutrons. Where are the latter coming from? The hydrogen fusion mechanism has been elucidated and looks like this:

$$H^1 + H^1 \rightarrow D^2 + \beta^+ + \nu,$$

where H^1 represents a proton, D^2 is a deuterium nucleus composed of a proton and a neutron, β^+ is a positive electron (also called antielectron or positron), and ν is a neutrino. This means that in the reaction, one proton is conserved and the other one is converted into a neutron by emission of a positron and a neutrino (through the action of the weak force). The superscripts represent the number of baryons present in each nucleus.

Then the reaction proceeds as follows:

$$D^2 + H^1 \rightarrow He^3 + \gamma,$$

where He^3 is the light isotope of helium, containing two protons and one neutron, and γ is electromagnetic radiation of very high frequency. In the next step two of these helium nuclei fuse:

$$He^3 + He^3 \rightarrow He^4 + 2H^1,$$

where He^4 is the isotope of helium containing two protons and two neutrons. This series of reactions is called the pp-chain. In other words, stars fuse hydrogen to make helium, and large amounts of energy are also produced. The fusion reaction has two important consequences: first, the energy provided by the reaction stabilizes the star and prevents gravitational collapse under its own weight. Second, it is expected that, as a star ages, its H:He ratio will decrease, as hydrogen is being consumed to synthesize helium. Scientists have actually observed this process. The helium content of the Sun's core, for example, is very far from the primeval H:He ratio of 76 percent to 24 percent. The Sun, which is about 5 billion years old, contains 38 percent H and 62 percent He in its core. Another chain of reactions, called the carbon cycle, also powers stars, and it, too, results in the fusion of hydrogen into helium with release of energy. It is called the carbon cycle because its initial step is to fuse one carbon nucleus with one hydrogen nucleus to make a nitrogen nucleus. Carbon is regenerated at a later stage and cycled back into the chain.

In fact, nuclear fusion in stars does not stop at the formation of helium. It has been shown that all elements up to iron are synthesized in the core of stars. This happens after the star has exhausted its hydrogen and starts using helium as nuclear fuel. This process is made possible by the gravitational contraction of the helium core, which brings its temperature to 100 million K and allows helium fusion. The first step is the fusion of three helium nuclei to make carbon. Through a complicated chain of reactions, more helium nuclei fuse with the newly synthesized carbon to form heavier elements, such as neon, sodium, and magnesium, until iron is reached. Iron, which contains twenty-six protons (and a variable number of neutrons, depending on the isotope) is the most stable element and can be fused with other nuclei only under temperature conditions not found in normal stars. However, iron is certainly not the heaviest possible element; elements containing up to 103 protons have been identified. Where do these come from? To understand their origin, one must first learn about the fate of stars.

Our Sun is about 5 billion years old and has enough fuel left to shine for another 8 billion. However, it will not remain the same for the rest of its life. As hydrogen is being exhausted, and as helium fusion starts proceeding, the inner core of a star heats up, as we saw earlier. One consequence of this higher core temperature is the expansion of the outer shell of the star, which becomes cooler. Such stars are called red giants because their diameter expands enormously and their cooler outer layer emits light in the less energetic red region of the spectrum. The Sun will eventually become a red supergiant, engulfing and incinerating Earth. When fusion in the core of a red supergiant stops, the star collapses and becomes a white dwarf, whose fate is to become an extinguished star once all fusion reactions stop.

Stars that are significantly more massive than the Sun experience a very different fate. When such a star reaches the end of its red supergiant state, its core collapse, because of its higher mass, is extremely violent and quick. Instead of a white dwarf, the core is turned into an extremely dense neutron star (where protons are fused with electrons) or an even more dense black hole (so called because the density and gravity of such an object are so high that even light cannot escape it). This violent collapse causes a shock wave that makes the outer layers of the star explode and blow out into space (figure 2.2). This phenomenon is called a supernova. The shock wave brings the star's outer layers to extremely high temperature, high enough to force atomic nuclei to fuse and synthesize elements heavier than iron. (This is a great simplification of the actual mechanisms leading to the synthesis of heavy elements.)

Thus this is where chemical elements other than hydrogen and helium come from: they are made in stars and released into space through the violent death

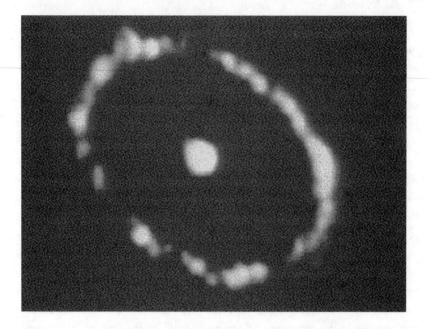

FIGURE 2.2 Ring of ejected stellar debris surrounding the remnant of a star that became a supernova in 1987 in the Large Magellanic Cloud. The ring is expanding at 6 million miles per hour. Photo taken by the Hubble Space Telescope. (Courtesy of the National Aeronautics and Space Administration.)

of the stars. The amount of matter released into space by a supernova explosion is numbered in the billions of billions of billions of tons (about 10^{27} tons or more). For reference, the mass of Earth is only about 6×10^{21} tons. Therefore a single supernova explosion releases into space the equivalent of 1,000,000 Earths or more. Supernova events are not rare; they occur about once every 100 years in the average galaxy. It is this star dust that is at the origin of planets and second-generation stars—that is, stars formed from the remnants of supernova explosions mixed with primeval hydrogen and helium. The Sun is such a second-generation star.

Interestingly, deep space observations with the Hubble space telescope (named after the discoverer of the galactic redshift-distance correlation) have revealed that the galaxies at high redshift, as far away from Earth as 10 billion light-years, do not show the complex chemistry observed in nearer galaxies. This is because these galaxies were formed only 2 billion years after the Big Bang and have not yet had time (as we see them today) to produce much in terms of elements heavier than hydrogen and helium. The existence of life in these galaxies (again, as we see them today) is therefore unlikely.

First-generation stars are those that were created directly from the primeval hydrogen and helium clouds. These stars are called iron-poor because they contain less iron than another class of stars, called second-generation stars, which contain more iron (and many other elements) than first-generation stars because they originate from the debris of supernova explosions. This debris, containing gases, metals, and silicates in the form of dust, forms visible clouds in the universe (figure 2.3), and these clouds are the birthplace of second-generation stars. These clouds can collapse and concentrate, perhaps through the shock wave of a nearby supernova, to form rotating, further collapsing discs. As the center of the disc reaches fusion ignition temperature, the second-generation star lights up. However, this does not mean that the whole dust disc is involved in star formation. In fact, a good portion of the disc remains separated from the newborn star and can actually be seen as an opaque halo around the star. Astronomers have detected this type of structure around several nearby stars (figure 2.4); it is called a protoplanetary disc because it is thought that planets originate from the coalescence of the outer regions of the disc.

In addition to H, He, metals, and silicates, a protoplanetary disc contains molecules commonly found in space, such as molecular hydrogen (H_2), water (H_2O), nitrogen (N_2), hydrogen cyanide (HCN), carbon monoxide (CO), and ammonia (NH_3), which can all be identified by their specific spectra. These compounds, together with many metals, are found in comets that sometimes go under the familiar name of "dirty snowballs." Comets are thought to be composed of the primordial matter that formed the protoplanetary disc that at one time surrounded the Sun. In the case of the solar system, the protoplanetary disc broke up into ten rings, representing the nine future planets and the asteroid belt found between Mars and Jupiter. According to one model, the heat of the Sun liquefied the frozen gases, with the result that the grains of protoplanetary matter became moist and agglomerated (accreted) into chunks of material. These chunks, called planetisimals, started colliding with one another through gravitational attraction and accordingly grew larger, forming the planets and their satellites. Computer simulations show that this process of accretion may have taken 100 million years. This view was, however, challenged recently by the observation of star and planet formation in the Orion nebula (visible with the naked eye as the middle "star" of Orion's knife). These tele-

FIGURE 2.3 Dust clouds with star formation: the Eagle Nebula. Photo taken by the Hubble Space Telescope. (Courtesy of the National Aeronautics and Space Administration.)

scope observations indicate that planets could form in as little time as 10 million years.

The energy stored by accreting planetisimals (the result of the kinetic energy possessed by colliding planetisimals) must have been considerable and, combined with the heat of decaying radioactive elements present in them, may have liquefied the growing planets. As a result, the metals, which are more dense than silicates, sank to the center of the newly formed planets, while silicates and other minerals floated toward the surface. At the same time, the planets degassed and acquired their original atmospheres consisting perhaps of hydrogen,

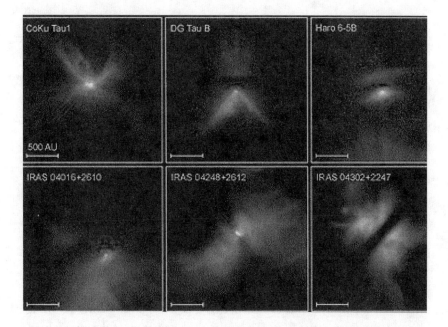

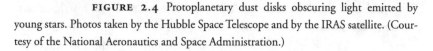

FIGURE 2.4 Protoplanetary dust disks obscuring light emitted by young stars. Photos taken by the Hubble Space Telescope and by the IRAS satellite. (Courtesy of the National Aeronautics and Space Administration.)

helium, methane, water vapor, nitrogen, ammonia, and hydrogen sulfide (plus small amounts of the inert gases argon, neon, krypton, and xenon). Some of these gases were generated through volcanic activity and the others corresponded to primeval galactic gases. Note that there was no oxygen as gas (O_2) anywhere in the solar system. The small planetary bodies, like the Moon and Mercury, quickly lost their atmospheres, while the larger ones retained them. The process of planet formation in the solar system stopped 3.8 billion years ago, as shown by the study of lunar rocks that still hold the traces of the heavy bombardment period that corresponded to the end of planetary accretion.

Until recently there was no evidence that planetary systems exist around stars other than the Sun. This view has changed considerably since 1995, the date of the discovery of the first extrasolar planet in orbit around the star 51 Pegasi. At the end of the year 2000, about four dozen extrasolar planets had been discovered. Our present instruments can detect only large extrasolar planets, the size of our Jupiter (the largest planet in the solar system) or larger (figure 2.5). Quite unexpectedly, several of these very large extrasolar planets have orbits around their stars that are smaller than Earth's orbit around the Sun. These planets have thus been dubbed "hot Jupiters." We do not know the composition of these planets, but what we do know is that our theory of planet formation may have to be revised once we learn more about them. Indeed, until 1995, the only planetary system we had been able to study was our own, and in our case, smaller planets like Mercury, Venus, Earth, and Mars are closest to the Sun. The gas giants—Jupiter, Saturn, Uranus, and Neptune—have much larger orbits. We do not yet understand how the hot Jupiters could have formed and remain stable so close to their stars in the extrasolar systems. This not to say that hot Jupiters are gas giants; we simply have no information on them yet.

The formation of planets begs the question of how life can appear on them. In other words, under what conditions is a planetary body able to generate life? Since the only life we know is based on carbon, hydrogen, oxygen, nitrogen, phosphorus, sulfur, and a few other minor elements, these elements must be present on the planet. This should not be a problem with planets revolving around second-generation stars. Then there is the question of the atmosphere. This problem will be discussed in chapter 4. For now let us just say that there is general agreement on the absence of oxygen gas in the atmosphere of the young Earth. This should not have been an impediment to the appearance of life. Indeed, we know today of many living species that do not require oxygen for proliferation and for whom, in fact, oxygen is very toxic. Hence, oxygen in the atmosphere is not a prerequisite for life. Other gases such as methane, carbon monoxide, carbon dioxide, ammonia, and hydrogen sulfide can perfectly well sustain certain life-forms, as we will see in the next chapter.

For life to exist, liquid water should be present, at least part of the time. We know of life forms deeply embedded in Antarctic ice, but these cells still depend on the percolation of liquid surface water between ice crystals. The same holds true for cells living deep inside Earth's crust. The presence of liquid water directly depends on the planet's distance from its star, hence its surface temperature and also its mass. A planet too close to the Sun, such as Venus, cannot have liquid water because it is too hot, while the Moon's gravity is too low to hold any kind of gas (including water vapor, if it was ever present). Mars may have had abundant water in its distant, warmer past, but this water is now locked into polar caps, where it sublimates rather than melts and is eventually lost to space as a result of low gravity and decomposition by ultraviolet light. The gas giants of the solar system have atmospheres composed mainly of hydrogen and

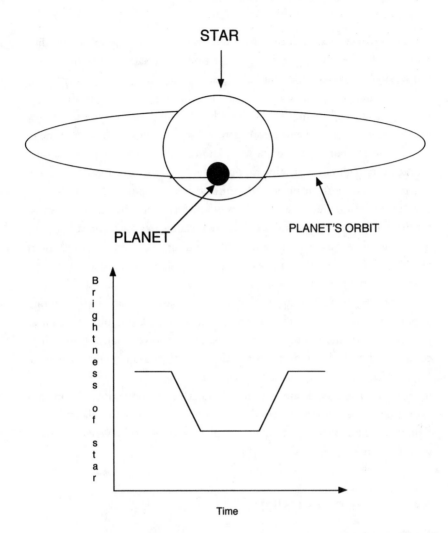

FIGURE 2.5 **Top:** Representation of an extrasolar planet eclipsing its star. **Bottom:** This graph shows the drop in the star's brightness as the planet transits in front of it. The star is the sun-like HD 209458. (Adapted from Doyle, L. R., H.-J. Deeg, and T. M. Brown. 2000. Searching for shadows of other Earths. *Scientific American* 283:58–65.)

helium. Methane, water, hydrogen sulfide, ammonia, and nitrogen are also present, making their atmospheric chemistry a potentially interesting one. Not enough of this chemistry is known, including temperature gradients in the at-mospheres of these gas giants, to preclude the existence of life, or precursors thereof, or to disprove it.

Thus a planet must meet certain orbital characteristics to be able to host life. Suitable orbits may vary greatly, depending on the surface temperature of the star itself and the age of the star. Old, first-generation stars may not have any accompanying planets at all because of a lack of heavy elements in the proto-stars that generated them. On the other hand, young stars may not have any planets either, because their protoplanetary discs have not had enough time to accrete. Furthermore, massive hot stars, known as blue supergiants, burn their fuel at a very high rate and shine for only a few million years. In all likelihood, this time is too short for planets to form and life to appear. Stars with masses close to that of the Sun or less (such as red dwarfs) can sustain nuclear fusion for many billions of years and are thus good candidates for planetary systems harboring life. A planet hospitable to life orbiting a relatively cool red dwarf would have a smaller orbit than a corresponding planet orbiting a yellow, hot-ter star like the Sun. Finally, about two thirds of the stars exist as binary or higher-order systems. This means that these stars have companions, all orbiting a common center of mass. The mutual gravitational pulls of multiple stars may prevent the formation of planets from dust discs or force putative planets into complicated orbits bringing them too close and then too far from their stars. Nevertheless, it has been calculated that, depending on the masses of the stars in the system and their distance from each other, orbits compatible with life can exist. If the appearance of life is a direct consequence of star and planet forma-tion, it is not impossible that millions of planets in the Milky Way are hosts to some form of life.

CONCLUSIONS

At its beginning, the universe was a very small object, much smaller than the size of a single proton. This object, however, contained an enormous amount of energy. Possibly through a quantum fluctuation allowed by the uncertainty principle, this object inflated at tremendous velocity for a very small fraction of a second, during which all four forces of nature appeared. Following this rapid inflation, the rate of expansion decreased and subatomic particles were created, as soon as the decreasing temperature of the nascent uni-verse allowed it. Protons (hydrogen nuclei) stabilized 1/100,000 of a second after the Big Bang, and at 100 seconds, the nuclei of the second element made in the Big Bang, helium, stabilized. It took an extra 400,000 years for matter to cease to exist as a plasma, the time at which electrons were captured by atomic nuclei. By then the radius of the universe had reached about 1/100,000 of its present size.

Clouds of primeval hydrogen and helium condensed into stars perhaps as early as 100 million years after the Big Bang, and stars started synthesizing all other elements found today in the universe. The Sun, a second-generation star originating from reprocessed supernova debris, formed about 9 billion years after the Big Bang, and its protoplanetary disc then started accreting into planets and their satellites. Atmospheres were generated through hot (at least plastic, perhaps liquid) planet degassing in the case of planetary bodies close enough to their star. At the end of the heavy bombardment period, about 3.8 billion years ago, Earth was ready for life. But first, to understand the past, we must study the present and know what life is and how it works. It is now time to leave the realm of physics and cosmology and enter that of biology.

CHAPTER 3

Life as It Is Today

Nothing in biology makes sense except in the light of evolution.
—T. DOBZHANSKY

I t is often said that biology is reducible to chemistry and that chemistry is reducible to physics. Quantum mechanics and thermodynamics have certainly shaped chemistry into the precise and versatile science we know today. However, saying that biology is merely complicated chemistry is going too far. If this were true, we could predict all the properties of an organism simply by computing all the quantum states, under a variety of conditions, of all the atoms composing the organism. This is clearly not (yet) possible. We cannot do it because of the sheer complexity of living systems. Thus a totally reductionistic approach to life is not helpful.

Consequently, this chapter will appear more descriptive and less theoretical than the two previous ones. This is by necessity. Biology is not yet (and perhaps never will be) a theoretical science like physics. Biology consists mostly of a very large collection of empirical observations, concepts, and definitions. It is just this multitude of biological phenomena that constitute living processes. Any theory (or theories) of the origins of life must account for these phenomena. So, in effect, the descriptions we encounter in this chapter will serve as the tests for the theories we meet in subsequent chapters. Nevertheless, the life sciences are not just a hodgepodge of unrelated facts. Life scientists have made great progress by using a combination of physical, chemical, and genetic techniques to uncover many common threads uniting all life-forms. Two major themes developed in this chapter concern the nature of genes and their expression, and their evolution. Like the universe, the gene is not a static entity; in a sense, it is also in constant motion.

Living organisms come in an amazing variety of shapes, sizes, and colors. At first sight, it is hard to determine what an octopus, a butterfly, a cat, and even microscopic algae share in common. Yet all these living creatures' lives are based on the same type of blueprint composed of the elements carbon, hydrogen, nitrogen, oxygen, and phosphorus. This blueprint is their genes and their genes are made of deoxyribonucleic acid (DNA), itself part of a class of molecules called nucleic acids because they were first found in the part of the cell called the nucleus (not to be confused with an atomic nucleus). Ribonucleic acid (RNA) is the second and only other member of this class. Life is not possible without DNA because it is DNA that stores in its molecule all the information needed to make an octopus an octopus, a butterfly a butterfly, and so on. To understand the origins of life, one must understand the origin of DNA. Before getting into the discussion of the origin of DNA, however, it is important to understand how genetic information is stored in this molecule and how this information is turned into the properties that various organisms display and pass on to their progeny. This topic is at the core of the science of genetics.

THE UNIVERSAL BLUEPRINT

The fact that DNA contains the information that tells an organism to be what it is (bacterium, oak tree, or human being) and what it does (be a parasite, grow leaves, or think) was discovered only in 1944 by Oswald Avery and coworkers at Rockefeller University. The science of molecular genetics, which deals with hereditary mechanisms at the molecular level, is thus quite young—several years younger than nuclear physics, for example. Yet enormous progress has been made in our understanding of genes in the past five and a half decades. It is now very well established that all life-forms on Earth rely on DNA-based heredity,[1] and that the mechanism through which hereditary traits are expressed is essentially the same across all species. The basic questions raised by molecular genetics are thus these: how is it that DNA stores the necessary information to determine what an organism looks like and does, and second, how is this information translated into shapes, colors, organs, movement, metabolism, and so forth?

Let us first look at the information storage problem. DNA is in fact a fairly simple molecule; its chemical constituents are four different nitrogen-containing bases [adenine (A), guanine (G), cytosine (C), and thymine (T)], each bound to the sugar deoxyribose. These base-deoxyribose combinations are linked together by phosphate groups to form a long, stringlike molecule. This molecule is double stranded, meaning that there are two strings bound together through chemical interactions between the bases, themselves located on the inside of the molecule. This molecule constitutes the famous double helix (figure 3.1), discovered in England in 1953 by James Watson, Francis Crick, Rosalind Franklin, and Maurice Wilkins. DNA molecules are extremely long, measuring from about 1 millimeter, in bacterial cells, up to several centimeters in human cells. These measurements tell us that DNA is very tightly packed inside cells whose sizes are measured in millionths of a meter. A close look at the DNA molecule reveals great uniformity. First, the pairs of interacting bases have the same width throughout. Second, the deoxyribose-phosphate group combinations that form the backbone of DNA are set up in a continuous helical way up and down the molecule. How can such a smooth molecule contain the thousands of instructions necessary to determine the properties of the thousands upon thousands of different species that exist on Earth?

The secret lies in the sequence of the base pairs and the number of base pairs present in the DNA of the different species. Let us consider a DNA molecule containing 1 million base pairs and calculate how many different combinations of these base pairs can be produced by arranging them randomly. The number is a staggering $10^{600,000}$, meaning the number 1 followed by 600,000 zeros! Now 1 million base pairs is not a very large number for DNA found in nature. It corresponds to the size of DNA present in the simplest bacteria. Considering that DNA from vertebrates (for example, mice and humans) contains several billion base pairs, one cannot even start to imagine the astronomical potential for genetic diversity present in this seemingly simple molecule. Of course, the billions of billions of billions (and so on) of possible DNA molecules mentioned earlier do not exist in nature; the number simply represents the *potential* for genetic diversity, not necessarily the number of different DNA molecules actually existing in our world. Indeed, we will see later that natural selection has strongly restricted the types of genomes (the genome is the suite of all genes in an organism) that can survive on planet Earth.

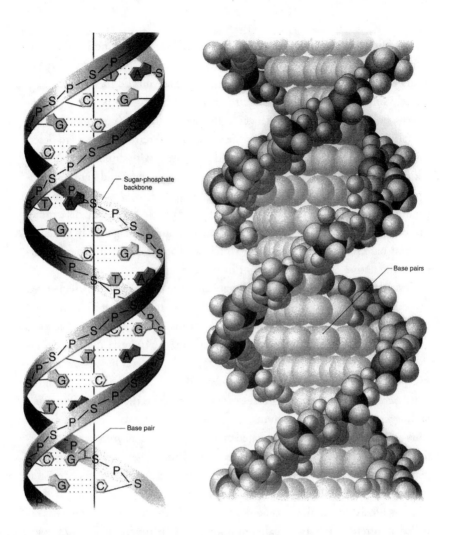

FIGURE 3.1 Two models of DNA. The drawing on the **left** shows the sugar-phosphate backbone (—S—P—) and the base pairs in the center of the double helix. The drawing on the **right** is a space-filling model in which the various atoms are represented by spheres proportional to their size. (Adapted from Hartwell, L. H., L. Hood, M. L. Goldberg, A. E. Reynolds, L. M. Silver and R. C. Veres. 2000. *Genetics*. New York: McGraw-Hill.)

In addition to looking at the sheer numbers of base pairs found in DNA, one must take into account the sequence of these base pairs. DNA base pairs are not arranged randomly in different species. For example, your DNA sequence is extremely similar to mine, because you and I both belong to the species *Homo sapiens*. For sure, there are differences in the details that make us different individuals, but the general blueprint is the same. If extraterrestrials were to determine the base sequence of your DNA and compare it with mine (without even taking a look at us), they would definitely conclude that you and I are very similar. More similar to one another than, say, either of us is to a lion. In fact, genome length (how many base pairs there are in DNA) generally determines the complexity of an organism—that is, the number of genes it contains, whereas base sequence (the order in which the base pairs are arranged) determines the nature of the genes present in this organism.

Several genomes from free-living organisms have been completely sequenced and their gene numbers tallied. (We are ignoring for the moment the many viral genomes—viruses are not free-living organisms—that have also been sequenced; see note 1 and chapter 5.) For example, the bacteria *Hemophilus influenzae* and *Escherichia coli* contain 1,830,137 and 4,639,221 base pairs, respectively, representing 1743 genes in the former and 4288 genes in the latter. Baker's yeast, *Saccharomyces cerevisiae*, possesses 6000 genes, and the fruit fly *Drosophila melanogaster* has 16,000. The human genome contains in excess of 3 billion base pairs and its gene number is estimated at about 35,000. This number will be refined once the human genome is fully sequenced and its genes precisely located, perhaps in 2003. (Only two independently obtained rough sequences exist at the time of this writing.)

A detailed study of all the genomes that have been sequenced to date has revealed an interesting point: we do not understand the function of 40 to 50 percent of the genes in these genomes. (The exception is the bacterium *Mycoplasma genitalium*, whose genome, the smallest sequenced so far, has only 580,070 base pairs representing 470 genes; only 20 percent of these genes have no known function.) The known genes (the ones whose functions we *do* understand), in cells from the simplest bacteria to the fruit fly and even humans, tend to be homologous. This simply means that the sequences of these genes, in spite of being found in evolutionarily very distant species, are quite similar, and these genes play similar roles in the cells of these species.

Undoubtedly, a significant proportion of the newly discovered genes will prove to be responsible for the individual characteristics of the various species.

For example, we do not expect to find genes for insect wing development in bacteria. Nevertheless, the fact that such a high proportion of homologous genes exists across living organisms very strongly pleads in favor of a common ancestry for all. Much has been said by opponents to evolutionary theory about the differences between the insect compound eye and the mammalian eye, often in an effort intended to deny any relationship at all between insects and mammals. Well, it turns out that the genes determining eye formation in the fruit fly and eye formation in the mouse are not that different. Basically, both sets of genes determine the formation of a light-sensitive organ, the "generic eye," with structural details filled out in a species-specific manner. The evolutionary relationship is unmistakable.

The unity of life is not only found in the similarities between DNA sequences among species. It can also be recognized in the complex machinery that transfers the genetic information encoded in the DNA to what is called the *phenotype* of cells and organisms. Simply put, the phenotype is the sum of all the properties that organisms display; these include visual appearance (with or without a microscope, depending on the organism's size), and the ability (or inability) to metabolize or manufacture certain compounds. For example, many bacteria and plants are able to manufacture vitamin B_1, whereas humans are not. The ability to resist antibiotics in the bacterial world and high or low grain yield in cereals are also phenotypic properties.

It turns out that all living cells and organisms use the same processes to transfer the genetic commands in DNA to the end of the line, the phenotypic properties of these cells and organisms. To understand this, one must first understand that most phenotypic properties are determined by proteins. Proteins are large molecules composed of amino acids chemically linked together. They come in an amazing variety of lengths, shapes, and amino acid compositions, which explains the enormous diversity of phenotypes encountered in nature. For example, whether one has dark skin or white skin depends on the amount of skin pigment one synthesizes. Pigments are manufactured through the action of biological catalysts called enzymes,[2] which are proteins. Whether one is lactose tolerant or lactose intolerant depends on the presence or absence of a functional protein-enzyme called lactase. Other proteins are not enzymes and are not involved in metabolism but play a structural role in cells. For example, collagen gives skin its elasticity, and keratin gives hair its threadlike appearance.

The information storage mechanism in DNA has been elucidated, starting in the 1960s by researchers such as Marshall Nirenberg and Gobind Khorana in the United States. All proteins are made of only twenty natural amino acids. Therefore if there is indeed a correspondence between the sequence of bases in DNA (the genetic information), which contains only four different informational elements; the four bases A, T, G, and C; and the amino acid content of proteins (the phenotype), which contain twenty different building blocks, one should be able to find a mathematical relationship between the DNA base content and the proteins' amino acid content.

Let us assume that it takes a single DNA base to determine one single amino acid in a protein. Then, since there are four different bases in DNA, only four different amino acids could be specified. This is not enough, since we know there are twenty different amino acids in proteins. What if the four DNA bases were arranged in groups of two in the DNA sequence, such as AT, AG, GT, CA, and so on? Would there be enough combinations to determine twenty amino acids? The answer is no, because four different bases arranged in groups of two (called doublets) offer only sixteen ($4^2 = 16$) different combinations, still short of the twenty needed. What about the four DNA bases arranged in blocks of three (called triplets), such as ATG, GCC, ACG, and so on? Would there be enough combinations to account for the twenty natural amino acids? The answer is yes, and more. The number of triplets one can make with four different bases is $4^3 = 64$! This number is well in excess of the twenty needed. In other words, doublets of bases do not provide enough combinations, but triplets do. Either the triplet hypothesis is right or nothing makes any sense.

The triplet model was proven right by laboratory experiments involving all sixty-four synthesized triplets and showing that sixty-one of them led to the specification of all twenty amino acids. Why were sixty-one triplets needed and what happened to the three triplets found not to specify any amino acids at all? Geneticists discovered that what is now called the genetic code (the ensemble of all sixty-four triplets, now called codons) is degenerate. This means that some amino acids can be specified by several different codons. For example, the amino acid leucine is specified by six different codons, TTA, TTG, CTT, CTC, CTA, and CTG. The amino acid lysine is specified by two different codons, AAA and AAG, whereas tryptophan is specified by a single codon, TGG. Three codons (TAA, TAG, and TGA) do not specify any amino acids. These are now known to be *stop* or *termination* codons, because they direct the system to stop inserting more amino acids into a growing protein chain and thus "decide" when a protein molecule is complete. The genetic code of sixty-four codons (figure 3.2) is universal (with very minor exceptions that are understood), mean-

Second letter

		T	C	A	G		
	T	TTT ⌉ Phe TTC ⌋ TTA ⌉ Leu TTG ⌋	TCT ⌉ TCC ⌉ Ser TCA TCG ⌋	TAT ⌉ Tyr TAC ⌋ TAA Stop TAG Stop	TGT ⌉ Cys TGC ⌋ TGA Stop TGG Trp	T C A G	
First	**C**	CTT ⌉ CTC ⌉ Leu CTA CTG ⌋	CCT ⌉ CCC ⌉ Pro CCA CCG ⌋	CAT ⌉ His CAC ⌋ CAA ⌉ Gln CAG ⌋	CGT ⌉ CGC ⌉ Arg CGA CGG ⌋	T C A G	**Third**
letter	**A**	ATT ⌉ ATC ⌉ Ile ATA ⌋ ATG Met	ACT ⌉ ACC ⌉ Thr ACA ACG ⌋	AAT ⌉ Asn AAC ⌋ AAA ⌉ Lys AAG ⌋	AGT ⌉ Ser AGC ⌋ AGA ⌉ Arg AGG ⌋	T C A G	**letter**
	G	GTT ⌉ GTC ⌉ Val GTA GTG ⌋	GCT ⌉ GCC ⌉ Ala GCA GCG ⌋	GAT ⌉ Asp GAC ⌋ GAA ⌉ Glu GAG ⌋	GGT ⌉ GGC ⌉ Gly GGA GGG ⌋	T C A G	

FIGURE 3.2 The universal genetic code. The first base position (on the left) is read down, the second position is read across, and the third position is read down. Amino acid names are abbreviated: Phe, phenylalanine; Leu, leucine; Ile, isoleucine; Met, methionine; Val, valine; Ser, serine; Pro, proline; Thr, threonine; Ala, alanine; Tyr, tyrosine; His, histidine; Gln, glutamine; Asn, asparagine; Lys, lysine; Asp, aspartic acid; Glu, glutamic acid; Cys, cysteine; Trp, tryptophan; Arg, arginine; Ser, serine; Gly, glycine.

ing that viruses, bacteria, plants, and animals all use that same code to produce their proteins. What a great proof that life is indeed one!

But that's not all. In addition to the existence of thousands of homologous genes and the existence of the universal genetic code, organisms use very similar processes to convert the DNA language of the genetic code (we can now call it the *genotype*) into the amino acid language of the phenotype, the proteins. As I said before, the question is one of information flow between the genotype (base-containing DNA) and the phenotype (amino acid-containing proteins). How does a living cell "know" how to arrange the amino acids of a protein in an exact way that follows the sequence of codons in a gene? This mechanism too has been unraveled and is basically the same in all life-forms.

This is how it works. First, the DNA is faithfully copied into an RNA molecule that contains exactly the same sequence of bases (and of course codons) as the DNA. RNA is very similar to DNA, except that it contains the sugar ribose (instead of deoxyribose), it is mostly single stranded instead of double stranded, and the base uracil replaces the base thymine. This RNA molecule is called a messenger RNA (mRNA) because it conveys the message (the codons) present in the DNA. Basically, DNA and RNA "speak" the same language, the language of codons. The process of synthesizing an mRNA molecule is called transcription. The transcribed mRNA molecule is then translated into protein language.

Translation is a complicated process achieved by two key elements, transfer RNAs (tRNAs) and ribosomes. Transfer RNAs are small RNA molecules (they are chains of seventy to eighty bases) that actually perform the translation of DNA (through its RNA intermediary) language into protein language. For this, they must be able to speak both languages. And indeed they do. The tRNAs are able to bind specific amino acids, and they also possess the ability to "read" codons in the mRNA by recognizing these codons with a sequence of three bases called the anticodon. In other words, tRNAs line up along the mRNA molecule through codon-anticodon interaction, and by doing so, they put the amino acids they carry in perfect alignment with one another, as they should be in the protein molecule.

This process of alignment of tRNAs and amino acids along the mRNA molecule is performed on a little cellular body called the ribosome. Ribosomes are composed of dozens of proteins attached to several RNA molecules called ribosomal RNAs (rRNAs). Ribosomes can be seen as the workbenches on which protein synthesis is performed. They keep the mRNA and the tRNAs carrying amino acids in close contact with one another and in the right order. It is also on the ribosome that adjacent amino acids are chemically linked together to form the finalized protein molecule. Transcription and translation are illustrated in figure 3.3.

Proteins are extremely diverse in their composition and length. This means that each gene (and each mRNA) coding for a given protein will have a specific base sequence (codon sequence) and a specific length. Long genes code for long proteins, and short genes code for short proteins. Interestingly, the sequences representing tRNAs and rRNAs are themselves encoded within the DNA. The process of transcription/translation is a universal principle, and it occurs very similarly in all life-forms, from bacteria to humans. It should be noted that both

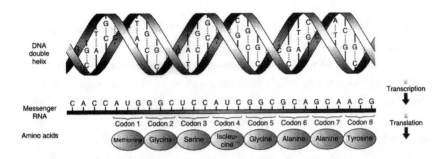

FIGURE 3.3 Information flow in living cells. A gene, represented by the DNA double helix, is first transcribed to generate a messenger RNA copy, and the copy is subsequently translated into a protein sequence. (Adapted from Alcamo, I. E. 1996. *DNA Technology: The Awesome Skill*. Dubuque, Iowa: Wm. C. Brown.)

transcription and translation rely on protein enzymes. The enzyme that performs transcription is called RNA polymerase and the enzymes that attach amino acids to tRNAs are called aminoacyl-tRNA synthetases. Without these protein enzymes, the system cannot function.

In terms of the origin of life, this poses an interesting conundrum: which came first, proteins or nucleic acids (either DNA or RNA)? Without protein enzymes, it is impossible to make RNA (and also DNA), and without genes, it is impossible to make proteins! We will see in chapter 4 some possible answers that have been put forth. Also, why is the process of transcription necessary? Why has life not evolved a system whereby the genetic information stored in DNA can be directly translated into protein language, without a need for mRNA? We will also see in chapter 4 that in fact, the first repository of genetic information may have been RNA, not DNA. These RNA molecules were probably very short, containing perhaps as few as 100 bases. The informational diversity afforded by such short nucleic acids is still 10^{60}, the number 1 followed by 60 zeros, still a staggeringly high number. More about that later.

One process I have not yet touched upon is the transmission of DNA-encoded hereditary traits to progeny cells or organisms. This process, too, is universal and based on the principle of DNA replication. We have already seen that DNA is a double-stranded molecule with interacting base pairs in its middle. These base interactions are very specific: an A can form a pair only with a T, while a C can pair up only with a G. This leads us to the notion that the two DNA strands are complementary. This complementarity is key to understanding how the genetic code is transferred from one cell to its daughter cells upon division. When cells divide, they first replicate (double) their DNA. To do this, there exists a complicated system, again based on protein-enzymes, that first separates the two DNA strands. Then, another enzyme, called DNA polymerase, copies each original DNA strand and generates the complementary strand of each by inserting the proper bases in the proper location. For example, when DNA polymerase encounters a G in the strand that it is copying, it will insert a C in the newly forming strand. Similarly, an A will lead to the insertion of a T in the growing strand. At the end of the process, one observes two DNA molecules (each containing two strands) where originally there was only one. What is more, because of the A-T and G-C rule, the two daughter DNA molecules are absolutely identical to the original molecule (figure 3.4)! This is why, even after thousands of cell divisions, daughter cells contain the same genetic information as the original cell.

How do we know that all of this is true? Countless experiments have supported the DNA replication, transcription-translation, and universal genetic code theory. But it is perhaps the modern gene cloning technology that gives the most brilliant and definitive demonstration through which scientists now understand a lot about the mechanisms of life. If it is indeed true that the genetic code is universal and that transcription and translation do operate in basically the same way in all cell types, it should be possible to express, say, human genes in bacteria and bacterial genes in plants, right? And yes, this is possible. The human gene coding for insulin has been cloned (isolated and purified) from human cells and expressed in bacteria and yeast. Insulin for diabetics is now conveniently produced in these microorganisms. Similarly, corn plants have been engineered for insect resistance with a gene isolated from the bacterium *Bacillus thuringiensis*. The ears from these plants are now adorning the shelves of our supermarkets (for details, see my book *High Tech Harvest*). There are hundreds of other such examples, all pointing in the same direction: for better or for worse, genes from any species can be expressed in any other species through a technology we call genetic engineering. Our ability to shuttle genes back and forth between unrelated hosts proves again that life is one.

VARIATIONS IN THE BLUEPRINT

Now we all know that once in a while, an organism displays unexpected phenotypic properties. For example, some bacteria in a large popula-

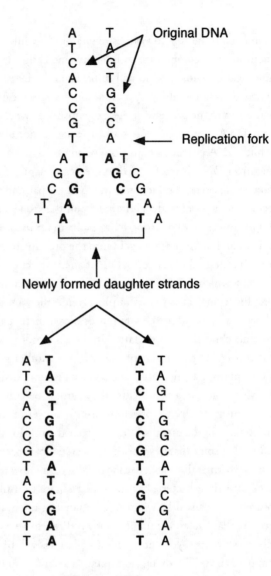

FIGURE 3.4 How DNA replicates. **Top:** The two strands of the double helix separate, and each strand is copied separately. **Bottom:** The result of replication is the production of two double helices with sequences identical to that of the original DNA molecule. Newly inserted bases are in bold type.

tion cease to be able to synthesize vitamin B_I while all the others go on manufacturing this compound. In humans, one child may be unable to metabolize the sugar galactose and so will suffer from the hereditary disease galactosemia. These defects are caused by a mutation in a specific gene. This mutation is then passed on to the progeny of these organisms. How can this happen? It turns out that DNA replication as described earlier is not absolutely perfect. In a rare while, DNA polymerase inserts a wrong base in a growing DNA strand.[3] This incorrect DNA is transmitted to the descendants of the cell in which the mistake has occurred, and it continues to be propagated as an incorrect DNA molecule. Since proteins depend directly on the sequence of codons present in the DNA, insertion of a wrong base results in an incorrect codon and thus insertion of an incorrect amino acid into the protein coded for by that gene. This is what a mutation is: the formation of an incorrect gene resulting in an incorrect protein, which, if it is an enzyme, has a great likelihood of not performing a correct function. In the two examples given earlier, a mutation occurred in one of the genes involved in the vitamin B_I biosynthetic pathway, whereas the galactosemic child carried a mutation in the gene coding for the enzyme responsible for converting galactose-1-phosphate into another galactose derivative. Galactose-1-phosphate then accumulates and becomes toxic.

Mutations are a fact of life; they cannot be completely avoided, since the mechanism of DNA replication itself is not error-proof. Most mutations are deleterious to the organisms that harbor them. Most, but not all. You probably know that some diseases caused by bacterial pathogens, which used to be easily curable with antibiotics a decade or so ago, are now resistant to treatment. This is because these bacteria have become resistant to antibiotics. In some cases, this resistance is caused by a simple spontaneous mutation in their DNA. Now if you are that bacterial mutant, you will thrive in an environment laced with antibiotics, whereas your nonmutant colleagues will become extinguished. What happens next? The dying nonmutant bacteria stop consuming food (present, for example, in the human intestine) and clear the way for the mutants to proliferate wildly since they no longer have any competition for resources. The result is uncontrolled multiplication of pathogenic mutants causing the patient to become very sick in spite of antibiotic treatment. From the standpoint of the mutants, nothing could be better: they have no competitors for food, they proliferate abundantly, and they are spread in nature—in humans, by diarrhea or coughing, for example. In other words, from a tiny minority, the mutants have become an overwhelming majority. This is great success in the struggle for life.

Thus some mutations can be beneficial to those (here, the bacteria) that harbor them.

This leads us directly to the concept of selection, critical in the understanding of evolution in general and evolution of early life-forms in particular. Selection consists in amplifying (making more numerous) organisms that otherwise would be present in a given environment in very small numbers. In the preceding example, the selective agent was the antibiotic that wiped out benign bacteria in the human victim, thereby creating a new niche for the pathogenic mutants to occupy. To put it slightly differently, the bacterial mutants display greater fitness in the antibiotic-laden environment than the nonmutants. Greater fitness means greater proliferation. In this example, the selective agent is an antibiotic, a human-made chemical. We call this process artificial selection. Other examples of artificial selection are higher yield in crop plants and milk production in cows, both brought about by selective breeding implemented by human agents.

It is not difficult to imagine a similar situation occurring in nature, without antibiotics and without intervention of breeders. Populations of living creatures are genetically heterogeneous. This genetic diversity is clearly visible in humans and though not so visible in a herd of buffaloes, the genetic variation is there (one notable exception is cheetahs, which are so inbred as to make them almost clones of one another; this is why they are in danger of extinction). In other words, large populations are collections of individuals sharing genes that make them belong to this or that species, but there exist many slight variants for many of these genes, making each individual unique.

When we look at wild populations that have occupied their ecological niches for thousands of years, we know that the individuals composing these populations are fit to survive and reproduce successfully in that particular ecological niche. This means that the genetic variation found in this population is well adapted to the environmental conditions of the niche. Now suppose that these environmental conditions change. This change can be catastrophic or gradual, like a sudden drought or a subtle climate change. It is at this level that genetic variation will play a large role; those individuals that, through subtle genetic differences already in place, possess better resistance to the new environmental threats, will be better fitted to the new conditions than their partners and will become more successful competitors for resources.

Over time (millions or billions of years), much genetic variation can be accumulated by living populations, and many ecological situations can change. This leads to the appearance of new species and the extinction of others. In a nutshell, this is how natural selection and evolution take place. The key to understanding evolution is that genetic variation preexists in populations. The variants (it is better not to call them mutants to avoid the risk of negative connotation) may represent a small minority under current environmental conditions, but they can quickly become a majority if these conditions change in a way that makes the variants better fit than the majority. The force at work (through, for example, climate change, fire, flood, asteroids falling, or volcanic activity) is natural selection. It is natural selection that has molded the different species, hence the different genotypes and their accompanying phenotypes that exist on Earth.

Not all of the innumerable and possible DNA base combinations exist in the biosphere because a very large number of them would be incompatible with the conditions that have existed and now exist on the planet. In fact, there is strong and fascinating evidence from phylogeny that all living things are descended from a common ancestor that lived prior to 3.5 billion years ago.

The branch of genetics called phylogeny busies itself with the genetic relationships that exist within and among species. For a long time, phylogeny was mostly based on the phenotypes of organisms, and it was concerned with characteristics such as shape, skeletal features (including those of fossils), and organ structure. Then, when it became possible to determine the amino acid sequence of proteins, in the 1950s, molecular technology became prevalent. The invention of DNA sequencing in the late 1970s made comparisons at the level of genes a reality.

It all started with the study of two proteins, cytochrome c and hemoglobin, in various mammalian lineages. Researchers discovered that the number of amino acid substitutions in these two proteins was proportional to the evolutionary distance separating them. For example, there is only one amino acid change when human and monkey cytochrome c are compared. There are, however, twelve such changes when human and dog cytochrome c are compared. This number increases to sixty-six in a comparison between humans and yeast. Furthermore, the number of substitutions is only one in a comparison between the horse and the donkey.

The fossil record, too, shows that humans and apes are more closely related than humans and dogs. Similarly, we know that horses and donkeys are very closely related. Thus a comparison between the fossil record and the rate of amino acid substitutions in proteins gave scientists the idea that such substitutions could be used as a molecular clock. The more changes there are between two similar proteins found in different species, the more distantly related they are. Similar experiments with other proteins gave an excellent correlation between the percent amino acid substitution and time since various species diverged, based on fossil evidence. The fossil record can thus be used to calibrate the molecular clock. Fossils can indeed be dated with great accuracy by a variety of techniques, some based on radioactive decay.

Since changes in amino acid composition are the result of changes in the DNA codons, DNA sequencing can also be used to study the relatedness between species. Studies at the DNA level have confirmed the results obtained with proteins (figure 3.5). Measuring the evolutionary distance based on DNA sequence divergence between extant organisms allows scientists to perform what has been called molecular archaeology. This is accomplished by reversing the procedure described in the previous paragraph. There, fossil evidence was used to calibrate the molecular clock. Then why not use that calibrated molecular clock to calculate the time at which various species diverged? This would be extremely useful in those cases where fossils do not exist, such as with most of the predecessors of the thousands of microbial species that exist today.

This was indeed done, and it has led to the construction of what are called phylogenetic trees. Figure 3.6 shows the general phylogenetic tree of life with its three domains, the Bacteria, the Archaea, and the Eukarya. Bacteria and Archaea are all microscopic, single-celled organisms whose DNA is not enclosed by a membrane. They are known collectively as prokaryotes. In contrast, Eukarya (also called eukaryotes) comprise single-celled and multicellular organisms, from yeast, to lettuce, to humans. Their DNA is always found in a membranous body called the nucleus. Scientists built the phylogenetic tree linking the three domains of life using comparisons between thousands of protein and DNA sequences. In this tree, the lengths of the various branches are proportional to the length of time since divergence. For example, the long branch uniting the Eukarya with the Archaea is about 1.8 billion years long. This means two things: first, that Eukarya derive from Archaea, and second, that it took about 1.8 billion years for some Archaea to evolve into Eukarya. The tree shown here is rooted (the root is the thick vertical line), and this root represents the universal ancestor(s) from which all life on Earth is descended. Chapter 5 will discuss this root in detail and explore some of the hypotheses related to the evolutionary splits that occurred between the three domains of life as we know them today. But first, let us study another paradigm common to all life-forms: energy processing.

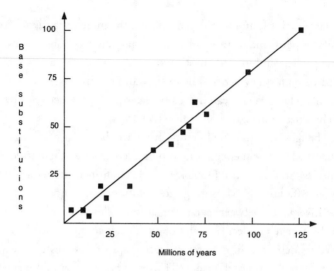

FIGURE 3.5 Calibration of the molecular clock. This graph represents the number of base substitutions in the genes of seven different proteins from seventeen species of mammals as a function of the evolutionary time separating these species. It can be seen that the relationship is linear: the more distance in the past, the more base changes. This graph indicates that the rate of base substitution for these genes taken together is 0.41 per million years. (Adapted from Ayala, F. J. 1982. *Population and Evolutionary Genetics.* Menlo Park, Calif.: Benjamin Cummings.)

METABOLISM AND ENERGY TRANSACTIONS

The thousands of living species that inhabit our planet may seem entirely disparate in the way they survive and reproduce. We have seen so far in this chapter that all life is based on DNA and built on essentially the same blueprint. Yet we cannot escape thinking that cacti, for example, must handle one of the processes of life, metabolism, very differently from say, tigers. Here again, however, there are more similarities than differences. It is natural to object to such a sweeping statement, because, after all, cacti are green and do not move, whereas tigers are not green and do move. Certainly, cacti possess some metabolic pathways that tigers do not, and vice versa, but remember that what keeps them both alive is matter and energy exchange between their cells and the outside world. Differences between types of organisms should not be downplayed, however, and chapter 5 will make clear that we are quite ignorant about

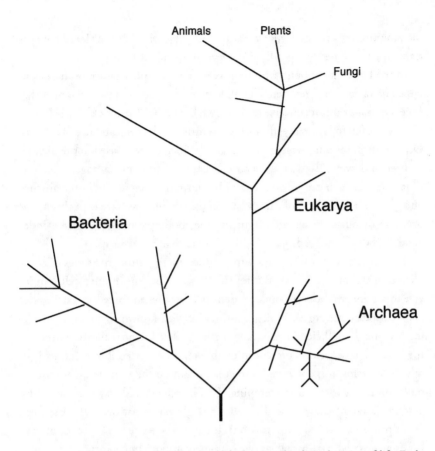

FIGURE 3.6 The tree of life showing the three domains of life. Each domain contains several branches (most of them unnamed in this drawing, but known) that correspond to different organisms. Archaea and Bacteria are all microorganisms devoid of nucleus and grouped under the name *prokaryotes*. The domain Eukarya is composed of unicellular and multicellular organisms, all equipped with a cell nucleus where DNA is concentrated. Eukarya are also known by the name *eukaryotes*. Divergence between Archaea and Bacteria took place 2 billion years ago, and that between Archaea and Eukarya, 1.8 billion years ago. Fungi, plants, and animals diverged from other Eukarya about 1 billion years ago. *Homo sapiens* (not represented at the scale of this diagram, but included under Animals) diverged from other hominids about 200,000 years ago.

the evolutionary origins of many features of living cells. For now let us concentrate on what unites living cells at the level of metabolism.

In the beginning, about 3.5 billion years ago, our planet's atmosphere contained no oxygen gas. Yet there is good fossil evidence that life was extant at that time. Humans cannot conceive of their own life in the absence of breathable air. However, even today, many life-forms can thrive in the total absence of this gas, O_2, so critical for our own survival. Examples include yeast used in the making of beer and wine, as well as nefarious bacterial human pathogens such as *Clostridium*, which causes gangrene or botulism. In a sense, life in the absence of oxygen is a "memory" of things past, and even humans have kept this memory: one of our own energy-producing metabolic pathways is oxygen independent. We share this property with all other living organisms.

The oxygen we breathe is used to produce energy through a series of complicated metabolic pathways. But for short periods of time, human cells can revert to an ancestral oxygen-independent pathway when forced to exist under anoxic conditions. For example, in a situation well known to athletes, breathing becomes less efficient and oxygen intake depleted. Under those conditions, the sugar glucose, always present in the blood, is converted into lactic acid (in other organisms, it is converted into ethanol and carbon dioxide), whose accumulation in muscles causes cramping. This metabolic pathway is catalyzed by a dozen different enzymes and is called the glycolytic pathway (the last steps of the process are called fermentation). It functions in the total absence of oxygen. The glycolytic pathway has one extremely important characteristic: it generates two molecules of adenosine triphosphate (ATP) per molecule of glucose metabolized.

What is the function of this ATP? Many, many chemical reactions connected with the synthesis of cellular components (such as DNA and proteins) require an energy source. The universal energy donor is ATP, thanks to the presence of two energy-rich phosphoanhydride bonds (figure 3.7). The breakage of these bonds, which occurs when ATP is involved in a cellular chemical reaction, releases energy that can be used to drive other reactions. Therefore glycolysis can be seen as the degradation of glucose to produce energy that can be used elsewhere in the cell. And again, this process can take place under anaerobic conditions, meaning in the absence of oxygen. Thus even humans possess today an ancestral metabolic pathway that does not require oxygen to produce energy. Glycolysis, however, is only one of several pathways that can produce ATP and hence energy.

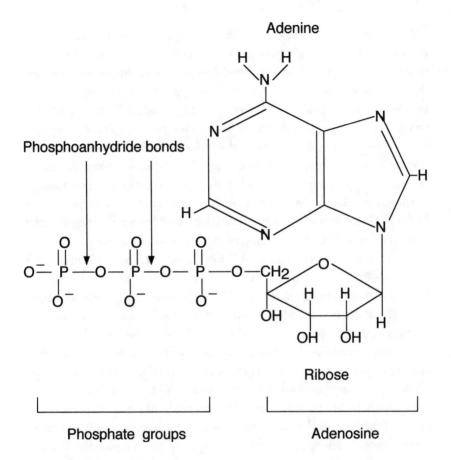

FIGURE 3.7 The formula of ATP. Adenine is a nitrogenous base (also found in DNA and RNA), and adenosine is the combination of adenine with the sugar ribose. AMP, ADP, and ATP are produced by successive additions of phosphate groups to the adenosine part.

We will see in chapter 5 how our atmosphere acquired its oxygen, but for Now suffice it to say that this oxygen was produced by early prokaryotes able to perform photosynthesis. Only plants and photosynthetic microorganisms can perform photosynthesis, because this process requires pigments (some of them called chlorophylls and carotenoids) that animals and many other microorganisms do not possess. Yet here also, one of the main functions of photosynthesis, in addition to producing oxygen and sugars, is ATP synthesis, the making of the fuel of life.

Photosynthetic pigments are present in cellular bodies called chloroplasts and are organized into light-harvesting complexes. The function of these complexes is to capture the photons of light, delivered for free by the Sun, and turn this light energy into chemical energy in the form of ATP. The initial stage of photosynthesis is the cleavage of water into oxygen, which is released into the atmosphere, and positively charged hydrogen ions (protons). The term *ion* simply refers to atoms or groups of atoms that have gained or lost one or several electrons. Since hydrogen contains a single electron, its loss turns the neutral atom into a single positively charged proton. These protons then flow through an enzymatic complex that generates ATP. It is estimated that one molecule of ATP is synthesized for every three protons that travel through the complex. And again, the fate of this ATP is to power other reactions taking place in the cell.

Finally, photosynthetic and nonphotosynthetic eukaryotic organisms that can metabolize oxygen (they live under aerobic conditions) have developed a third ATP-generating system called aerobic respiration. This process takes place in small cellular bodies called mitochondria (singular: mitochondrion) that probably appeared about 2 billion years ago. Many prokaryotes also perform aerobic respiration, but they do not contain mitochondria. Aerobic respiration depends on one of the products of glycolysis, pyruvic acid, which is further metabolized inside the mitochondria into carbon dioxide, which is released into the atmosphere. Electrons generated in these reactions perform an important function: they cause a flow of protons (always present in the aqueous environment of the cell) through an enzymatic complex that, when energized by these protons, makes ATP. Furthermore, these protons react with oxygen and reduce it to water molecules. Therefore the result of aerobic respiration is to convert glucose into carbon dioxide and water, the process being accompanied by the synthesis of thirty-six to thirty-eight ATP molecules. Aerobic respiration is thus a much more efficient way to produce energy than glycolysis, which produces only two ATP molecules per each metabolized glucose molecule.

In conclusion, metabolic processes in all living cells are made possible thanks to the existence of the universal energy donor, ATP. Various organisms have evolved three different strategies to generate ATP: glycolysis, photosynthesis, and respiration. Glycolysis makes possible the existence of life in the absence of oxygen, photosynthesis makes use of the light energy from the Sun, and aerobic respiration makes use of oxygen, a gas that represents 21 percent of our atmosphere. As we understand it, glycolysis evolved first, followed by photosynthesis and eventually aerobic respiration. Plants are the ultimate ATP producers: they synthesize it via all three pathways. Animals, being devoid of chloroplasts, can generate ATP only by glycolysis and aerobic respiration.

Everywhere in the biosphere, ATP synthesis depends on coupled electron transfer reactions. In these reactions, a *reduced* donor transfers electrons to an *oxidized* acceptor. These oxidation-reduction reactions are universal, but one should take care not to misinterpret the term *oxidation*. In common language, oxidation means chemically attacked by oxygen. Chemistry uses a more general definition involving electron flow from one compound to another. Oxygen (O_2) is but one oxidized (obviously) electron acceptor; other examples used by some living cells are the ferric ion (Fe^{3+}), the nitrate ion (NO_3^-), and the sulfate ion (SO_4^{2-}). In effect, cells that use ferric ions, nitrate ions, or sulfate ions "breathe" these ions instead of oxygen. This type of metabolism is called anaerobic respiration. Electron donors are very often organic nutrients, but they can also be hydrogen gas (H_2), carbon monoxide (CO), or hydrogen sulfide (H_2S) (in the case of some bacterial species). In the final analysis, life as we know it today is possible because oxidation-reduction reactions exist.

Photosynthesis and aerobic and anaerobic respiration all ultimately depend on a class of molecules called porphyrins (figure 3.8). These molecules are pigments, often attached to a protein chain, that catalyze the transfer of electrons from the donor to the acceptor. This electron (e^-) transfer is often accompanied by the addition of protons, H^+, which are ubiquitous in water. A typical reduction reaction commonly performed in organic chemistry laboratories—but also occurring in living cells—can be written as follows:

$$R\text{-CHO} + 2e^- + 2H^+ \rightarrow R\text{-CH}_2\text{OH}.$$

This reaction reduces an aldehyde (R-CHO) to an alcohol. R is any chemical group, usually containing carbon atoms. It can be seen that the product of the reaction has gained two hydrogen atoms and is thus concomitantly reduced. In this example, the aldehyde is the electron acceptor, while the electron donor is not specified.

In aerobic respiration, the electron donor is pyruvic acid and the electron acceptor is oxygen, which is reduced to water. In living cells, the electron acceptor does not react directly with the electron donor. Rather, electron transfer occurs via a complicated pathway involving several intermediates containing porphyrins. In aerobic respiration, the electron transfer system consists of cytochromes, proteins attached to a porphyrin ring containing an iron ion at its center. In anaerobic respiration, where oxygen is absent, the electron donor can

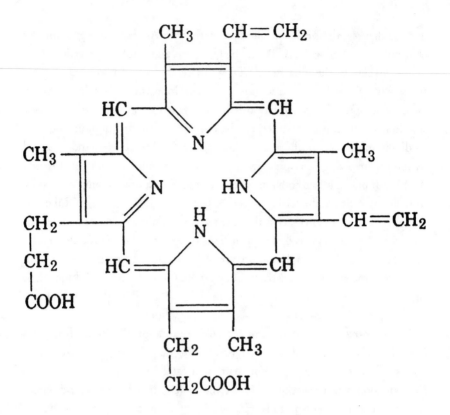

FIGURE 3.8 The general structure of porphyrins. In hemoglobin and myoglobin, the porphyrin ring contains an iron ion at its center. In chlorophyll, this ion is magnesium.

be sulfate, SO_4^{2-}, which is reduced to sulfide, S^{2-}, to be released either in the form of hydrogen sulfide gas, H_2S, or, if dissolved metals are present, in the form of precipitated sulfides. Sulfate-reducing bacteria are extremely ancient, as they have left their mark in minerals 3.4 billion years of age. Here also, cytochromes containing porphyrin rings are involved. And finally, all animals equipped with a circulatory system possess blood, whose characteristic color is caused by the presence of a porphyrin-containing molecule, hemoglobin (as it is called in the case of mammals, fish, birds, and reptiles). As we all know, the function of hemoglobin is to ferry oxygen and release it at the cellular level.

Photosynthetic organisms contain chlorophylls, a class of pigments equipped with a porphyrin group containing a magnesium ion at its center. Photons of visible light bump the electrons of the porphyrin ring to a higher energy level and, in the case of higher plants, these electrons are used to split water into protons and oxygen gas. The protons are then used to reduce the compound nicotinamide adenine dinucleotide phosphate (NADP) to its hydrogenated form, NADPH. NADPH is then used to drive the reactions that convert carbon dioxide into sugars (reducing CO_2 in the process), through a complicated mechanism that requires ATP. Some photosynthetic bacteria are able to perform photosynthesis without splitting water, and hence without producing oxygen. Here, the electron donor can be a sulfide (SH^-, such as H_2S) or a thiosulfate ($S_2O_3^{2-}$) but here also, NADP is reduced to NADPH. The interesting consequence of this process is that elemental sulfur, instead of oxygen, can be produced. Cytochromes are also involved in photosynthesis. Clearly, porphyrins found in chlorophylls and cytochromes play a key role in electron transport mechanisms and hence energy transactions in cells. Their origin must be understood. A highly simplified rendition of some oxidation-reduction reactions taking place in living cells and involving porphyrins is given in table 3.1.

LIFE THAT DEPENDS ON SIMPLE MOLECULES ONLY

Organisms that feed on complex organic molecules seem familiar to us. These organisms include animals and fungi. Their metabolism seems familiar because humans are part of this class of organisms, called heterotrophs. All animals find their complex nutrients in a diet of animal or plant products, or both. Therefore this type of life-form depends on preexisting life as a source of food. This does not hold true for photosynthetic organisms and many prokaryotic life-forms. Photosynthetic organisms, called phototrophs, do not depend on any outside electron donor (except water) or acceptor to satisfy their energy needs. As we have seen, the photons of visible light are converted into chemical energy (by electron transfer) via chlorophylls. Furthermore, the substrate they reduce is carbon dioxide, a molecule that is abundant in the atmosphere, which they reduce into sugars. Nitrogen, phosphorus, and necessary metal ions (such as magnesium and iron) are provided by the soil in the form of phosphate, nitrate, or ammonia and other salts. Therefore phototrophs depend for survival only on simple inorganic compounds that they eventually convert into proteins and nucleic acids.

TABLE 3.1 Some Energy-Producing Metabolic Pathways Involving Porphyrins

PATHWAY	ELECTRON/ HYDROGEN DONOR	ELECTRON/ HYDROGEN RECEPTOR	REACTION PRODUCTS
Anaerobic lactic acid fermentation	Lactic acid	A sulfate (SO_4^{2-}) such as Na_2SO	Acetic acid H_2O + a sulfide (S_2^-), such as H_2S
Aerobic respiration	Pyruvic acid	O_2	$CO_2 + H_2O$
Photosynthesis	H_2O	NADP	NADPH + O_2

Note: All reactions are coupled with the formation of ATP from ADP. In photosynthesis, NADPH powers the reduction of atmospheric CO_2 into glucose.

Interestingly, some of the ammonia used by phototrophs is produced by atmospheric nitrogen fixation, which occurs in some bacteria. These prokaryotes contain an enzyme, nitrogenase, that enables them to convert nitrogen gas (N_2), which constitutes about 79 percent of the Earth's atmosphere, into ammonia. This is done through the action of sulfur- and iron-containing proteins called ferredoxins that act as electron transfer elements. This reaction is also a reduction but porphyrins are not involved. Rather, the active center of ferredoxins consists of a lattice of interconnected sulfur and iron atoms. Cyanobacteria evolved 3.5 billion years ago and are the oldest known extant life-forms. They perform photosynthesis that produces oxygen, they perform nitrogen fixation, and they survive under anaerobic conditions. We will see in chapter 5 what this means for the evolution of life.

Finally, a class of prokaryotes called chemoautotrophs do not rely on photosynthesis but use simple inorganic compounds to produce their nucleic acids and proteins. Their high-energy electron donors are inorganic molecules such as hydrogen gas (H_2), hydrogen sulfide gas (H_2S), ferrous iron (Fe^{2+}), and even carbon monoxide (CO). They can reduce CO_2 into organic compounds, and in their niche, they find phosphorus and nitrogen in the form of dissolved inorganic salts.

All these observations demonstrate that life can thrive on minerals (ubiquitous on land and in the oceans even before life appeared) and sunlight, clearly present at the origin of the solar system. In a nutshell, life could have started from inorganic matter. Therefore the first attempts at primitive metabolism, occurring in water before the first cells were born, need not have used complicated organic molecules as precursors but could produce some of these molecules through electron transfer—that is, oxidation-reduction chemistry. What could those reactions possibly have been? This will be covered in chapter 4.

THE CELL'S ENVELOPE AND ITS SKELETON

All cells are bounded by an envelope that separates them from the outside world and from other cells. This envelope, the cell membrane, determines which compounds can and cannot penetrate the cells, which compounds should be excreted, and under what circumstances these events occur. When we compare Archaea, Bacteria, and Eukarya, we find many similarities but also some major differences in the nature of this envelope. Let us first consider the similarities.

All cell membranes are based on the same model: the lipid (mostly phospholipid, with some other lipids such as cholesterol) bilayer. Phospholipids are composed of two major elements: two hydrocarbon tails (fatty acids) and a head containing a phosphate group (PO_4^-) (figure 3.9). Because of its negative electrical charge, this phosphate group is hydrophilic, meaning that it is soluble in water. In contrast, the hydrocarbon tails are hydrophobic—they are insoluble in water. Molecules that have affinity for water but also repel it are called amphiphilic. When phospholipids are suspended in water, they spontaneously form microscopic spherical structures composed of two layers (the bilayers). This is because the phosphate-containing heads are miscible in water (present both outside and inside the spherical structure), whereas the hydrocarbon tails are not. The tails present in the two layers interact with one another to form a sheet (see figure 3.9). Phospholipid bilayers are quite fluid, meaning that they can easily fuse with one another to make bigger bilayers. On the other hand, they can be divided by fission—by hard shaking, for example. Vesicles composed of pure phospholipids are highly impermeable to water and water-soluble molecules and ions. Living cells are clearly *not* impermeable to water and the compounds dissolved in it, such as nutrients, and this is because cell membranes also contain special proteins that act as portals to let some compounds in and others out. This influx-outflux activity is highly regulated in both prokaryotes and eukaryotes.

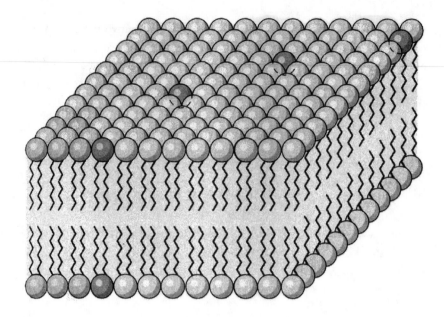

FIGURE 3.9 The phospholipid bilayer. The spheres represent nega-
tively charged phosphate groups that interact with water. The zigzag lines represent the hydrocarbon "tails" of the phospholipid molecules that line up with one another and are present inside the bilayer.

Now what about the differences between the three domains of life? First,
many cell types possess a cell wall in addition to a cell membrane. The cell wall
is used for structural containment. Among the Eukarya, animal cells do not
have a cell wall, but plant cells and fungal cells do. In plant cells, the cell wall is
composed of polysaccharides (such as cellulose), whereas in fungal cells, a compound called chitin is used. Among prokaryotes, the bacterial cell wall is composed of peptidoglycans, whereas the archaeal cell wall is made of pseudopepti-
doglycans or protein. There are also differences in membrane composition:
eukaryal and bacterial fatty acids in the bilayer are attached to glycerol molecules via a type of chemical bond called an ester bond. In Archaea, this bond is
the very different ether bond. In addition, the archaeal lipid bilayer is much less
fluid than the eukaryal and bacterial bilayers because it contains long, two-
chained hydrocarbons that span the entire width of the bilayer and in a sense
freeze it in position. This sturdier arrangement of the archaeal bilayer may be

an adaptation to the harsh, high-temperature environment in which many Archaea live.

Besides an external cell membrane, Eukarya possess an intricate network of inner membranes. The nucleus, the compartment that contains most of the cell's DNA, is surrounded by two membranes, each consisting of a lipid bilayer plus proteins. There are holes in these membranes, nuclear pores, for import and export of proteins, small molecules, and RNA. In addition, animal and fungal cells contain mitochondria, the cellular organelles where ATP is synthesized. Mitochondria, too, have a double membrane, and so do chloroplasts, the plant organelles where photosynthesis and ATP synthesis take place. Interestingly, both mitochondria and chloroplasts possess their own DNA. We will see in chapter 5 what the DNA harbored by these organelles may mean in terms of their origin. Photosynthetic bacteria do not contain distinct chloroplasts. Their photosynthetic apparatus is either located in the cell membrane or in invaginations derived from it. Finally, Eukarya host other organelles, devoid of DNA, and bounded by a single lipid bilayer. These are lysosomes, peroxisomes, and glyoxysomes, all involved in digestive and detoxification functions.

Organelles are not the only eukaryal cellular structures containing membranes. A eukaryotic cell contains an involved network of internal membranes comprising the endoplasmic reticulum and the Golgi complex. Parts of this network communicate with the nuclear envelope and the cell membrane. This complex network, including organelles, is called the cytomembrane system (figure 3.10). We will see that its origin is far from being understood.

Finally, eukaryotes, but not prokaryotes, contain an internal support system composed of interconnected filaments and microtubules collectively known as the cytoskeleton. The cytoskeleton provides an architectural framework, but it also participates in cell movement through contraction and expansion. Major players in these movements are the motor protein molecules known as actins, dyneins, kinesins, and tubulins.

CONCLUSIONS

We have seen in this chapter that all life-forms are based on a set of common paradigms; the ultimate blueprint of life is DNA containing the genetic information. This genotype then transfers its information to the phenotype via a complicated mechanism (transcription/translation) involving several

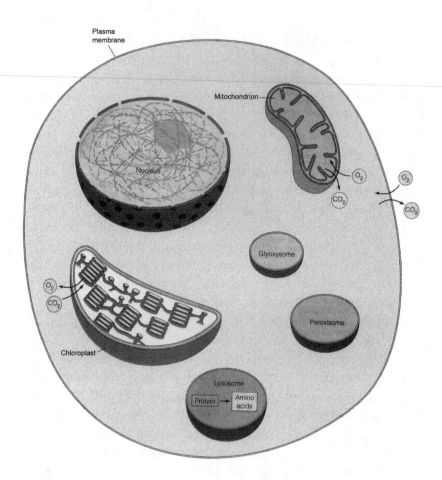

FIGURE 3.10 A composite generic plant/animal cell summarizing the features of eukaryotic cells. Glyoxysomes, peroxisomes, and lysosomes are organelles devoid of DNA that metabolize fatty acids, hydrogen peroxide, and proteins, respectively. Mitochondria generate ATP and CO_2, whereas chloroplasts generate ATP and O_2. The structures inside the chloroplast are grana, bodies where photosynthesis takes place. (Adapted from Becker, W. M., L. J. Kleinsmith, and J. Hardin. 2000. *The World of the Cell*. San Francisco: Benjamin Cummings.)

types of RNA molecules and protein enzymes. To understand the origins of life, one must understand the origin of DNA and the origin of transcription and translation.

Next, genes contain their own potential for change (mutation or variation) through base substitution. This mechanism thus turns populations of organisms into gene pools, in which some organisms are better adapted to new ecological situations. Natural selection then has the ability to change gene frequencies in a given direction and bring about new phenotypes, new species. DNA sequencing techniques now allow researchers to measure the evolutionary distance, in years, between different species. Clues about the origins of life are thus hidden within these DNA sequences and must be uncovered.

Moreover, we have seen that metabolic pathways lead to the chemical conversion of organic substances (such as glucose) into other compounds, the consequence of which is the release and storage of energy in ATP. ATP is then used to power many reactions that synthesize compounds such as RNA and DNA bases as well as amino acids. These building blocks are then used to make RNA, DNA, and proteins, reactions that also all require ATP to function.

All living cells are based on the same principles. Metabolic pathways are catalyzed by enzymes, and these enzymes are proteins (we will see in chapter 4 that some enzymes are made of RNA). To understand the origins of life, one must understand the origin of ATP (and its biosynthetic pathways based on oxidation-reduction reactions, including the role of porphyrins) and that of proteins.

Then all cell membranes are largely made of phospholipids organized in bilayers. These membranes regulate the import and export of substances into and out of cells. To understand the origins of life, one must understand the origin of membranes. The same holds true for the cytoskeleton, which exists in eukaryotes but not in prokaryotes. Where does it come from? And finally, our carbon-based life must have depended on the availability of organic molecules on the surface of Earth. Where did these come from? All these questions are addressed in the next two chapters.

CHAPTER 4

Prebiotic Earth:
First Organic Compounds
and First Informational Molecules

Where the hell do the likes of us come from, Hans Thomas?
Have you thought about that?

—JOSTEIN GAARDER, *The Solitaire Mystery* (1996)

As we saw in chapter 2, Earth's atmosphere 4 billion years ago was very different from the one we know today. There was no oxygen, but other gases were present. In one scenario, these were methane (CH_4), water vapor (H_2O), nitrogen (N_2), ammonia (NH_3), hydrogen sulfide (H_2S), and carbon dioxide (CO_2). Primeval hydrogen (H_2) and helium (He) were disappearing fast because Earth's gravity was not strong enough to keep them in the atmosphere. Traces of helium would always be present, however, thanks to the radioactive decay of elements, such as uranium, thorium, and radium, in Earth's interior. There were also oceans, whose geography we would not recognize today, since plate tectonics has moved the continents around. Volcanic activity contributed water, nitrogen, carbon dioxide, sulfur dioxide, and other gases to the atmosphere.

The sight must have been awesome: the sky was red and the Sun looked bluish because nitrogen, which scatters blue light, was not yet the dominant gas in the atmosphere. The atmospheric pressure was ten times higher than it is today. There was of course no life but there was movement; the wind blew clouds and volcanic smoke around while the waves lapped at beaches made of solid rock, not yet sand. High tides took place every three hours and the Moon appeared four times closer than today. The temperature was much higher than now, perhaps as high as 90°C, because of the greenhouse effect caused by atmospheric methane and carbon dioxide. Nevertheless, this temperature was low enough for liquid water to exist. In the absence of oxygen, there was no ozone (O_3) layer and Earth was bombarded by hard ultraviolet (UV) radiation from the Sun.

Scientists have gained much information concerning Earth's primitive atmosphere by studying our sister planet, Venus, whose atmosphere contains huge amounts of carbon dioxide (96 percent), some nitrogen (3 percent), and some sulfur dioxide (SO_2), but no oxygen. However, being closer to the Sun, Venus receives twice as much radiation and heat, which, combined with the enormous greenhouse effect caused by the vast amount of CO_2, raises the surface temperature to 460°C. This is where the parallels between Venus and Earth stop: liquid water cannot exist on Venus because it is too hot, so life is impossible on that planet. Mars has the opposite problem: being farther from the Sun, it is too cold for liquid water to exist. However, its present tenuous atmosphere is very much like Venus's, without the sulfur dioxide.

Conceivably, Venus and Earth originally had similar atmospheres, composed of, in order of decreasing abundance, H_2, He, CH_4, H_2O, N_2, NH_3, and H_2S. What happened to these gases on these planets? Today, Earth's atmosphere is composed of about 79 percent nitrogen, 21 percent oxygen, and low amounts of a few other gases. Venus also lost many of its original gases. As we have seen, H_2 and He quickly escaped into space. The fate of the other gases (except N_2) was sealed by the UV radiation from the Sun, which decomposed them through the phenomenon known as photolysis. On Venus, where liquid water did not exist, water vapor was quickly decomposed into hydrogen (which escaped) and free oxygen, which rapidly oxidized CH_4 and H_2S to CO_2 and SO_2. The oxidation of NH_3 simply produced more N_2.

On Earth, where liquid water did exist, H_2O photolysis was not nearly as pronounced because UV-induced water decomposition occurs best in the gas phase. Nevertheless, photolysis of the primeval gases did take place, albeit more slowly, because Earth is farther from the Sun. Thus CH_4, NH_3, and H_2S also gradually disappeared. There was one enormous difference between Earth and Venus, however: thanks to the presence of liquid water, the CO_2 and SO_2 generated by photolysis dissolved in the oceans, reacted with the minerals present there, and became locked in rocks. This prevented the runaway greenhouse effect that occurred on Venus and made nitrogen the dominant gas. After a few hundred million years, the sky finally turned blue. Before this happened—that is, while Earth's atmosphere was still reducing (still containing appreciable amounts of H_2, CH_4, and NH_3)—could it be that some of the building blocks of life were somehow assembled in the primeval gases? After all, Earth's crust and the oceans at that time could not have held much in terms of carbon- and nitrogen-containing compounds that make up living cells. Some scientists do indeed think that the first building blocks of life were synthesized via atmospheric chemistry.

ORGANIC COMPOUNDS FROM EARTH'S PUTATIVE PRIMITIVE ATMOSPHERE

In 1953, Stanley Miller, working in the laboratory of Harold Urey (the Nobel laureate who discovered deuterium) at the University of Chicago, published the results of a strange experiment that was based on the assumption that Earth's atmosphere was once a reducing one. In a reducing atmosphere, free oxygen does not exist, and elements are present in a reduced, hydrogenated form such as CH_4, H_2O, and NH_3, and, of course, H_2. Miller knew that the Russian Alexander Oparin and the British J. B. S. Haldane had suggested several decades earlier that a reducing atmosphere could have allowed the synthesis of organic compounds (composed of carbon, hydrogen, nitrogen, and a few other elements) necessary for life to appear. However, they never did any experiments to test that idea. Miller, then a graduate student, decided to test this hypothesis, apparently with only reluctant support from his boss, Urey (who did not coauthor the article).

Miller built an apparatus (figure 4.1), made of glass, into which he introduced liquid water, hydrogen, methane, and ammonia gases. (The air had previously been pumped out of the system to eliminate oxygen.) To produce water vapor, Miller boiled water held in a flask; this had the further effect of circulating the mixture of gases present in the apparatus. To simulate rain, Miller added a cooled condenser to the system and trapped the condensed water in a U-shaped tube. The condensed water thus represented the primitive ocean. Samples from this tube could be harvested and analyzed for any water-soluble molecules that formed. Finally, the apparatus contained another large glass flask, in which the gases circulated, and which housed two metal electrodes that allowed electrical sparking of the gas mixture. The electric discharge simulated the lightning that

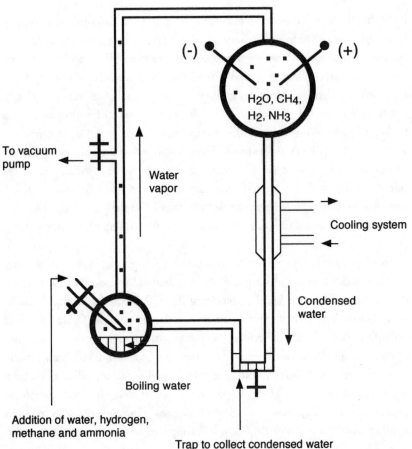

FIGURE 4.1 Schematic representation of the equipment used by Stanley Miller to produce organic molecules under abiotic conditions. The air from the glass apparatus is first evacuated by pumping. Water and the gases hydrogen, methane, and ammonia are then added to the flask at the lower left. Water is brought to a boil and water vapor starts circulating (*black dots*) in the apparatus. Water vapor and the three other gases then fill the large glass chamber at upper right. The chamber is equipped with two electrodes connected to a power source, and sparking is initiated. Water vapor and compounds resulting from the action of sparking are then condensed by the cooling system and accumulate in the trap. Compounds harvested from the trap are then analyzed chemically.

undoubtedly must have occurred in the atmosphere (the atmosphere of Jupiter is witness to huge lightning discharges and contains all the gases used by Miller).

After several days of cycling the gases and sparking, Miller noticed that the condensed water in the tube had turned pink and subsequently orange-red. Clearly, some chemistry was taking place, as the original gases were completely colorless. Analysis of the solution revealed the presence of amino acids, the building blocks of proteins! Of the twenty amino acids found in proteins, ten were formed in Miller's experiments. The chemistry that took place in these experiments is now understood. For example, the simple amino acid glycine results from the condensation of formaldehyde (formed from the sparked gases) with ammonia and hydrogen cyanide (also formed in the gas mixture) to produce the compound aminonitrile. Aminonitrile then reacts with water to form glycine (figure 4.2).

In addition to amino acids that make up proteins, gas-discharge experiments have also yielded the four nitrogenous bases, adenine (A), cytosine (C), guanine (G), and uracil (U), the building blocks of RNA. Adenine for example, results from the condensation of five molecules of hydrogen cyanide (figure 4.3). (It is unsettling to think that the poison used to execute prisoners in a gas chamber can lead to the synthesis of some of the building blocks of life!) Finally, many types of sugars were also synthesized in these experiments, including ribose, the sugar found in RNA. These sugars are made through polymerization of formaldehyde, itself formed in the sparked gases. However, ribose is made in small amounts, and it is unclear how it, as opposed to all other sugars, came to form RNA. We will see a possible answer to this problem later in the chapter.

The startling results of Miller's experiments have led to the notion that Earth's primitive oceans accumulated more and more of the building blocks of life, amino acids, nitrogenous bases, and sugars, and became some sort of primordial or prebiotic soup. (Instead of the term *soup*, which suggests a chunky mixture—think about split pea with ham or chicken noodle soup!—I prefer the word *broth*.) Primordial broths of the Miller type have been replicated by many investigators using similar gas mixtures exposed to short-wave UV light or silent electric discharge, all undoubtedly present on primitive Earth.

Interestingly, free oxygen completely interfered with all the reactions just described. Since we know that the oldest rocks found on Earth are not oxidized, this supports the conclusion that, indeed, oxygen was not present in the primeval atmosphere and could not have blocked the chemical reactions observed by Miller and others. Thus the problem seemed solved: the building blocks of pro-

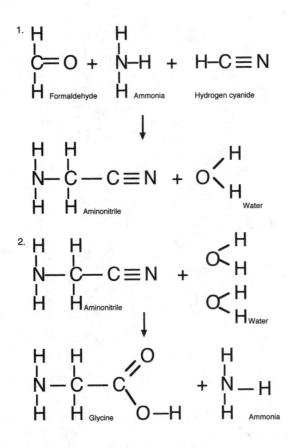

FIGURE 4.2 The two-step synthesis of glycine from formaldehyde, ammonia, and hydrogen cyanide in sparked gases.

teins and RNA (but not DNA, as thymine was not formed in these reactions) were made in Earth's reducing atmosphere.[1] But was the problem really solved? Perhaps not.

There are indeed difficulties with Miller's experiments. First, most scientists now concur that Earth's atmosphere did not contain significant amounts of hydrogen, methane, and ammonia for the length of time sufficient to allow the formation of a primordial broth based on these compounds. For these dissenters, the atmosphere quickly evolved into one containing mostly carbon dioxide and nitrogen (plus water vapor and unreactive gases such as argon and neon), much like the atmospheres of Venus and Mars today. Indeed, it is argued that Earth's primeval atmosphere (as well as those of Venus and Mars) was

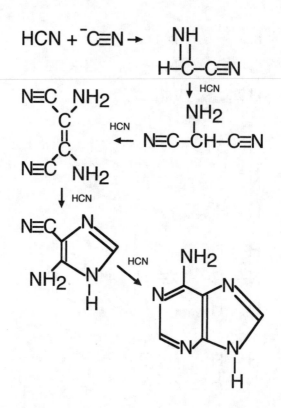

FIGURE 4.3 The synthesis of adenine from the condensation of six hydrogen cyanide molecules.

blown off as our Sun ignited. Another contributing factor would have been the violent collision between young Earth and a Mars-sized object that became the Moon. A new atmosphere could then have been generated through volcanic activity. It turns out that volcanoes emit mostly water vapor, carbon dioxide, and nitrogen, together with small amounts of carbon monoxide and hydrogen. No ammonia and methane are emitted by modern volcanoes. Therefore Earth's primitive atmosphere may have been much less reducing than imagined by Miller.

Furthermore, sparking mixtures of nitrogen, water, carbon dioxide, or carbon monoxide "à la Miller" gives somewhat disappointing results, albeit not entirely negative ones. Without hydrogen, CO_2 and N_2 cannot produce HCN, necessary for the formation of amino acids and nitrogenous bases. If, however, mixtures of CO_2, N_2, and H_2 or mixtures of CO, N_2, and H_2 are sparked, some amino acids, in reduced numbers and yields, are produced. Thus a strongly reducing atmosphere is necessary to produce the whole variety and large amounts of organic compounds found in a Miller-type of experiment. On the other hand, the nucleic acid base uracil could be produced by sparking mixtures of methane and nitrogen. Where was this methane coming from if it did not exist in the primeval atmosphere? It could have come from hydrothermal vents found on the bottom of the ocean or it could have been delivered by cometary impacts (see later). HCN, so crucial as an intermediate in the synthesis of amino acids and bases, could have originated from the reaction of methane with nitrogen to produce it and hydrogen. This reaction could have been catalyzed by UV light from the Sun. Hydrogen produced by this reaction, plus hydrogen emitted by volcanoes, would have made Earth's atmosphere mildly reducing. In conclusion, the atmosphere was probably not as reducing as Miller thought, but reduced carbon in the form of CH_4 as well as H_2 gas were probably present in modest amounts.

It was discovered in the early 1960s that simple mixtures of ammonia and hydrogen cyanide in water also form amino acids and large amounts of adenine. This suggested that bringing these three chemicals together somewhere, and not necessarily on Earth, could produce some of the building blocks of life. At about the same time, amino acids, as well as the bases adenine and guanine, were detected in meteorites.[1] Meteorites derive mostly from asteroids present between the orbits of Mars and Jupiter (some rare ones are of lunar or Martian origin). These meteorites were never exposed to any kind of atmosphere, reducing or oxidizing. How then do they contain organic compounds? These compounds would have had to have been made in space, possibly as a result of reactions between water, hydrogen cyanide, methane, and ammonia. Do these chemicals actually exist in space? Yes, they do.

SPACE CHEMISTRY AND
THE ORIGINS OF LIFE

We saw in chapter 2 that, indeed, large interstellar clouds in which stars and planets are formed do contain these molecules, plus many others. Water ice is thought to be the most common solid in the universe, and even methanol (CH_3OH; wood alcohol) and ethanol (C_2H_5OH; drinking alcohol) exist in deep space! What is more, these clouds also contain cyanoacetylene (HC_3N), a chemical that reacts with hydrogen cyanide to form nitrogenous bases found in RNA. Thus gases and particles left over from the protoplanetary disc after its condensation into planets could have delivered significant amounts of organic materials to Earth's surface in the form of meteorites and comets.

Comets in particular are thought to be aggregates of primordial material much predating the formation of the solar system as we know it today. They are found in two systems orbiting the Sun. The first system is called the Kuyper belt and its members orbit the Sun between the orbits of Neptune and Pluto. The second system, called the Oort cloud, forms a halo, well beyond the orbit of Pluto, around the solar system. Once in a while, a comet is knocked off its orbit because of the gravitational effect of the planets or a passing star. The comet then has a chance to approach the Sun at a much closer distance, with the result that some of its material evaporates and forms a visible tail. The tail can be studied and its spectrum recorded, either from Earth or from spacecraft on a rendezvous mission. Results of all studies agree that comets are rich in organic material and water. Comet Halley was even shown to contain the bases found in RNA and DNA. Comets visible from Earth are not that frequent today, but they may have been much more prevalent in the vicinity of Earth a few billion years ago when the solar system was forming. Collisions with young Earth may have been frequent and would have resulted in the seeding of the planet with building blocks of life or their precursors.

Similarly, meteorites, some of which contain organic compounds, as we have seen, still collide with Earth today and may have played the same role. It is known that planet Earth accumulates about 100 tons per day of meteoritic material in the form of micrometeorites. These micrometeorites measure about 0.1 mm in diameter and are carbon rich (up to 7 percent by weight). Micrometeorite samples have been collected in the "clean" environment of antarctic ice and have been shown to contain detectable amounts of organic molecules. It is quite possible that prebiotic Earth accumulated vast amounts of carbon-containing compounds by sweeping up micrometeoritic material.

If comets and meteorites and their cohort of organic molecules truly originate from primeval interstellar clouds (at a further stage of processing in the case of meteorites), these clouds should also contain molecules found in comets and meteorites. Astronomical observations have shown that this is the case. These clouds contain, in addition to gases, solid material in the form of dust. Their spectra indicate that this dust is made of microscopic silicate granules covered with ice composed of water, methanol, methane, carbon monoxide, carbon dioxide, and a class of organic molecules called polycyclic aromatic hydrocarbons. Given that interstellar clouds are constantly bathed in UV light emitted by nearby stars, the potential for interstellar chemistry is great.

Interstellar ice analogues have been created in the laboratory: the compounds known to exist in interstellar clouds were sprayed in a vacuum chamber kept at very low temperature to simulate conditions found in space. The ice grains thus formed were illuminated with UV light to mimic starlight. Analysis of these ices revealed that complex organic molecules could be formed, including quinones (electron transfer intermediates in photosynthesis), amino acids, and long-chain hydrocarbons with properties similar to amphiphilic lipids found in the membranes of living cells (see chapter 3). These amphiphilic hydrocarbons are able to form microscopic vesicles, presumably through bilayer formation, when dispersed in water. Intriguingly, similar compounds are found in some meteorites and they, too, form these vesicles in water. Such vesicles are able to trap a variety of chemicals, and by keeping these chemicals in close contact, they may have initiated some of the chemical reactions that led to life. Interestingly, amphiphilic hydrocarbons are not formed in Miller-type experiments. In conclusion, space contains a host of organic molecules able to generate, through chemical reactions, some compounds necessary for life to appear. Furthermore, space chemistry seems able to produce molecules not found in gases submitted to electric discharges.

Another objection to Miller's experiments is based on the chirality of the molecules he produced in his sparked gases. Chirality, also called handedness, refers to the three-dimensional arrangement of atoms in a molecule. A good analogy is that of the symmetry of human hands, hence the word *handedness*. The left hand and the right hand are directly superimposable when pressed palm to palm. They are not perfectly superimposable when pressed palm to knuckles. To achieve palm-to-palm contact, one of our hands must be rotated, or a single hand can be directly superimposed with its mirror image. The same thing happens in nature: amino acids come in two varieties, left-handed and right-handed, which can be distinguished by their ability to rotate polarized light either to the left or to the right. It turns out that amino acids in living cells belong to the left-handed category. One of the criticisms of Miller-type experiments is that they produce equal amounts of left-handed and right-handed amino acids, something not found in living systems. In contrast, amino acids found in meteorites contain slightly more left-handed amino acids than right-handed ones. This does not prove that life originated from meteorites, but it certainly does not rule out this notion.

Thus space itself, not necessarily Earth and its primitive atmosphere, may have been responsible for the synthesis of the building blocks of life. Or was it?

OCEAN FLOOR CHEMISTRY

As we saw in the introduction to this book, science constantly questions and requestions received knowledge. That some scientists think that organic compounds important for life were formed in the atmosphere or were brought to Earth by comets and meteorites does not make it so. Other hypotheses can always be formulated. A third hypothesis, which holds that the first organic compounds were made on Earth rather than in space or in the atmosphere, is the hydrothermal vents hypothesis.

Earth's crust is particularly thin at the level of ocean floors. There, tectonic plates slowly slide on top of the magma and by doing so, create ridges and cracks in the crust. Water percolates through these cracks, becomes superheated by contact with the magma, and dissolves many of the minerals present there. The pressure is so high that the water, heated to several hundred degrees, does not boil but is spewed back through chimney-like structures called hydrothermal vents (figure 4.4). These vents are often surrounded by miniature ecosystems, where many types of bacteria, worms, and crabs proliferate.

Now, high pressure and temperature can create some interesting chemistry. Could it be that the building blocks of life were once synthesized from purely inorganic compounds in or near hydrothermal vents? Theoretical studies performed in the mid 1990s indicated that, yes, this is possible. The branch of chemistry called thermodynamics mathematically determines what chemical reactions are possible or impossible. Using this approach, scientists realized that in water, the synthesis of all amino acids and their linking to form small proteins was theoretically possible above 100°C and under high pressure. The ingredients needed to achieve this synthesis were CO_2, ammonia salts, and H_2S (some amino acids contain sulfur). It turns out that both high temperature and these particular chemicals are found in oceanic hydrothermal vents. Therefore in theory, amino acids and proteins could have been formed at great depth under the surface of the ocean. Of course, such calculations did not prove that these reactions actually took place; some scientists used the derogatory term *paper chemistry* to characterize these studies. The hydrothermal vent hypothesis had to be put to the test.

This was done using an instrument called a bomb, which is simply a reinforced container able to sustain high temperatures and pressures. It was shown

FIGURE 4.4 An oceanic hydrothermal vent. (Courtesy of the National Oceanographic and Atmospheric Administration via the Smithsonian Institution.)

that, indeed, a whole catalog of organic molecules, including amino acids and pyruvic acid (an important metabolite ubiquitous in living cells), could be formed at high pressure and temperature from H_2S, CO, and CO_2 as well as ammonia (from nitrate) and nitrogenated hydrocarbons. Interestingly, iron sulfide (FeS) was absolutely necessary to catalyze these reactions, to generate hydrogen for reduction reactions, and to concentrate and stabilize the reaction products. This mineral is abundant in the earth's crust in the form of pyrrhotite. Although the formation of nitrogenous bases as found in RNA and DNA has not been reported, hydrothermal vent chemistry seems to be much more than just paper chemistry.

In summary, no one of the three hypotheses aimed at explaining the origin of the building blocks of life is complete. Experiments "à la Miller" have not yielded long-chain amphiphilic hydrocarbons like those found in cells, but amino acids and nitrogenous bases were produced. Space-based chemistry can produce the three types of compounds, but critics argue that organic molecules ferried by comets and meteorites colliding with Earth could not have survived atmospheric entry and impact because of the high heat generated. Finally, high temperature-high pressure chemistry, as is presumably taking place in hydrothermal vents, has so far not produced nitrogenous bases and long-chain lipids. How can this conundrum be solved?

It is always possible that more research on the preceding three systems will yield some new findings and solve the mystery. On the other hand, it is also possible that life did not need amino acids, nitrogenous bases, and long-chain lipids all at the same time. One can imagine, for instance, a situation in which genetic information (in the form of bases linked together to form a nucleic acid) came first, and that this genetic information started coding for some form of primitive metabolism. Or perhaps another scenario occurred, in which mineral catalysts or primitive enzymes, made of amino acids synthesized in hydrothermal vents or elsewhere, started a "protometabolism" that eventually led to the formation of nucleic acids. In that case, genetic information would have appeared as a result of a mineral-based protometabolism and, possibly, after the products it now codes for, proteins. Yet another possibility is that life appeared as a result of a complicated cooperation between simple proteins and simple nucleic acids. Scientists have not yet developed this third scenario. Therefore we will consider the "metabolism-first" world and the "RNA-first" world in that order, keeping in mind that these two models are not necessarily incompatible.

PROTEINS AND METABOLISM FIRST: THE IRON-SULFUR WORLD

Two of the most prominent proponents of the idea that protometabolism was key to the appearance of life on Earth are the Nobel Prize winner Christian de Duve of Belgium and Günter Wächtershäuser of Germany. Some of their work is discussed later. Both espouse the view that iron- and sulfur-containing compounds were critical to the establishment of protometabolism in the prebiotic world (a world still devoid of life as we know it but poised to see its birth). Such a view hinges on two ideas: first, a source of energy was needed to make possible some prebiotic chemical reactions, and second, the formation of proteins may have happened spontaneously.

As we saw in chapter 3, a great number of chemical reactions taking place in living cells require an energy supply in the form of ATP, a combination of adenine, ribose, and three phosphate groups. Before life existed, there was no abundant source of ATP. What could possibly have been the source of energy that drove prebiotic reactions in the primeval broth? In addition, assuming that the primitive Earth's atmosphere was not as reducing as thought by others, but that it contained carbon, mostly in the form of CO or CO_2 (but not CH_4), and no longer any hydrogen, how could hydrogenated (reduced) forms of carbon be produced? It is indeed mostly reduced carbon that is present in biological molecules. A possible answer lies in a combination of the effects of UV light from the Sun and the presence of iron on Earth's surface. This is how things may have worked according to de Duve. Not surprisingly, his model is based on oxidation-reduction reactions.

Iron, in the form of oxides and salts, is a very common element on Earth. When chemically combined with other elements, iron becomes ionized (electrically charged) and its ions can either be doubly positively charged (Fe^{2+}) or triply positively charged (Fe^{3+}). Fe^{2+} is the reduced form of iron, whereas Fe^{3+} is its oxidized form. Both forms currently exist naturally on Earth and must have existed in the distant past as well. Reduced iron is soluble in water, whereas oxidized iron combined with oxygen is not. Thus in the absence of oxygen, prebiotic Earth's waters must have contained significant amounts of reduced iron dissolved in water. Furthermore, water (H_2O) always contains a certain amount of protons (H^+) and hydroxyl radicals (OH^-). Remembering that oxidation-reduction reactions are accompanied by electron transfer (see chapter 3), it is easy to understand how the oxidation of iron may have powered some important chemical reactions on prebiotic Earth.

It turns out that under the influence of UV light (provided by the Sun), the Fe^{2+} ions can react with the protons present in water to form Fe^{3+} ions and hydrogen gas. In this reaction, reduced Fe^{2+} donates an electron to each proton (H^+) to give Fe^{3+} (an additional positive charge is gained because an electron is

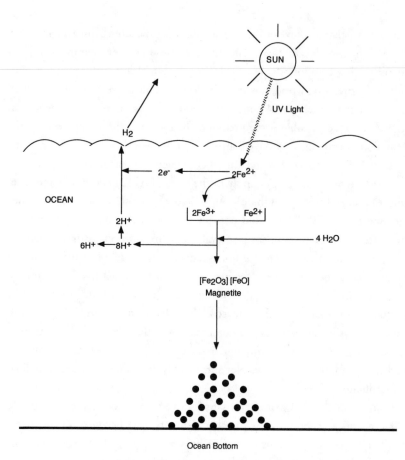

FIGURE 4.5 How the oxidation of iron ions in water irradiated by ultraviolet (UV) light produces hydrogen gas that escapes into the atmosphere. In this process, reduced iron Fe^{2+} is oxidized into Fe^{3+}, which reacts with water to produce insoluble magnetite. The electrons lost by Fe^{2+} in the oxidation reaction are captured by H^+ ions, always present in water, to give neutral hydrogen atoms that quickly combine to make molecular hydrogen gas, H_2. (Adapted from de Duve, C. 1991. *Blueprint for a Cell.* Burlington, N.C.: Neil Patterson.)

lost) and an H atom. Two H atoms quickly combine to give hydrogen gas, H_2, which escapes into the atmosphere. The Fe^{3+} so generated then reacts with water to give iron oxide, which precipitates out of solution as a black material. Thus iron is oxidized in this process while protons are reduced to hydrogen gas (figure 4.5). This hydrogen may have rendered the atmosphere more reducing.

Furthermore, and again under the influence of UV light, iron ions could possibly have cycled back and forth between a reduced and an oxidized state while reacting with CO_2. Carbon dioxide, a gas, is soluble in water. Therefore dissolved Fe^{2+} ions could have donated electrons to CO_2, which, in the presence of H^+, would have been reduced to CH_4 (methane) or $=CH_2$ compounds free to react with other substances. In addition, some of the newly formed $=CH2$ compounds could have spontaneously reacted with Fe^{3+} ions (before they precipitated) to regenerate Fe^{2+}. The interesting thing here is that this last reaction releases energy that could have been used to drive other reactions. This mechanism is called the iron cycle (figure 4.6).

Thus iron chemistry allows for the formation of reduced carbon compounds, including methane, hydrogen, and possibly other reduced chemicals such as ammonia, hydrogen cyanide, and hydrogen sulfide. It also produces energy in the absence of ATP. This scenario shows that a native and strongly reducing primitive atmosphere may not have been needed; the iron cycle produced the necessary reduced ingredients in solution and these could have combined to form amino acids and nitrogenous bases to make up the prebiotic broth. On the other hand, if the primitive atmosphere was reducing, the iron cycle could have provided energy to achieve reactions involving precursor compounds made in the atmosphere. Therefore the iron cycle and the formation of organic molecules in the atmosphere may have been synergistic.

Let us now examine how the iron cycle and its energy production could have been taken over by reactions that approximate living mechanisms (which the iron cycle does not). We saw in chapter 3 that the universal energy donor in living cells is ATP. Remember that it is the cleavage of one or two phosphate groups from ATP that generates energy usable by other reactions. Could the prebiotic broth have stored energy in compounds containing phosphorus but no adenine and no ribose (assuming these two were either absent or present in very low concentrations)? Yes, and the following reactions explain how.

We must first assume that a class of molecules called carboxylic acids existed in the prebiotic broth. These organic acids have the general formula R—COOH, where R is a group that may be simply a hydrogen atom, or it may be a methyl group ($—CH_3$) or a more complicated group of atoms.[2] It turns out that carboxylic acids are made abundantly in a sparked mixture of methane, water, hydrogen, and ammonia. They are also found, along with amino acids, in some meteorites. Then we must assume that another class of molecules, called thiols, also existed in the prebiotic broth. *Thiol* derives from the Greek word for sulfur, and the general formula of a thiol is R′—SH, where R′ can be

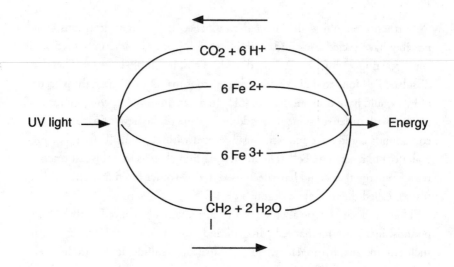

FIGURE 4.6 The iron cycle. As Fe^{2+} is being oxidized into Fe^{3+}, CO_2 dissolved in water is reduced into $=CH_2$ compounds (hydrocarbons) that can be used as organic building blocks. The reoxidation of some of these $=CH_2$ compounds into CO_2, coupled with the reduction of Fe^{3+} back to Fe^{2+}, produces energy that can be used to drive other prebiotic reactions. The cycle is catalyzed by UV light. (Adapted from de Duve, C. 1991. *Blueprint for a Cell.* Burlington, N.C.: Neil Patterson.)

H (in which case the thiol is simply H_2S, hydrogen sulfide) or a group containing reduced carbon, such as $—CH_3$, to make methyl thiol. We have seen that H_2S was probably originally present in the atmosphere of prebiotic Earth and if not, it was produced by hydrothermal vents. It is also soluble in water. It could thus have easily reacted in solution, under the action of UV light, with reduced carbon formed from carbon dioxide through the iron cycle, to produce a variety of organic thiols.

Next, the carboxylic acids could have reacted with the thiols to form a class of compounds named thioesters:

$$R—COOH + R'—SH \rightarrow R'—S\sim CO—R + H_2O$$

The $\sim$ sign in the thioester formula represents an energy-rich chemical bond. The energy necessary to form that thioester bond could have come from the iron cycle, or else this reaction could have occurred spontaneously at high temperature and high acidity (perhaps as in hydrothermal vents?). Finally, we must assume that phosphate ($H_2PO_4^-$) was present. Thus thioesters could have reacted with phosphate in the following way:

$$R'—S\sim CO—R + H_2PO_4^- \rightarrow R'—SH + R—CO—O\sim P—HO_3^-$$

and

$$R—CO—O\sim P—HO_3^- + H_2PO_4^- \rightarrow R—COOH + H_2P_2O_7^{2-},$$

where $H_2P_2O_7^{2-}$ is pyrophosphate. Pyrophosphate, like ATP, is an energy-rich, phosphate-containing molecule, and it could have played a role in protometabolism in the absence of adenine and ribose. What is also remarkable about this set of reactions is that the carboxylic acid (R—COOH) and the thiol (R'—SH) are regenerated and can thus enter a new cycle of energy-rich pyrophosphate production. Taken together, the iron cycle and thioesters provide an interesting scenario for energy transactions in the prebiotic broth. The author of this scenario, de Duve, has called it the "iron-thioester world." Needless to say, not everyone agrees with this scheme; particularly uncomfortable with it are those who think that protometabolism is a consequence of the appearance of genetic information, not the other way around.

Nevertheless, the iron-sulfur world has more to offer. Having tackled the energy problem, proponents of this view also think that protein enzymes may have appeared without genetic information. We saw in chapter 3 that proteins are made by the chemical linking of amino acids. This linking does not occur spontaneously and is performed in cells, as we have seen, through the mechanism of translation, involving not only the amino acids but also ribosomes and transfer RNAs. Assuming that ribosomes and transfer RNAs did not exist in the iron-sulfur world, how could proteins have formed? Again, thioesters come to the rescue.

As their name indicates, amino acids are acids. Not only that, they are also carboxylic acids containing the —COOH group that we encountered before. Thus amino acids could have reacted with thiols (R'—SH) to form thioesters, too. Furthermore, it has been demonstrated that amino acid thioesters can polymerize (make long chains of linked amino acids) spontaneously, without ribosomes and tRNAs! Therefore the prebiotic broth may have contained a large, random set of different proteins, all potentially composed of many different amino acid sequences. And these proteins, much like today's proteins made by living organisms, could have possessed enzymatic activity. These "protoenzymes" could have progressively synthesized the building blocks of nucleic acids (the nitrogenous bases), and they even could have diversified the pool of amino acids to form other types of protoenzymes. These many types of protoenzymes could then have been able to catalyze the synthesis of RNA or DNA—that is, genes. At this point, the prebiotic broth would have become informational.

There are, of course, problems with this view. For example, we do not know whether random chains formed from amino acid thioesters have any kind of significant and relevant enzymatic activity. This must be tested in the laboratory. Then it is unclear whether these putative protoenzymes (with an assumed catalytic activity) would have been able to coordinate their activities to produce any significant amount of genetic material, RNA or DNA. Thus the iron-sulfur world is an interesting, falsifiable hypothesis that must be buttressed by a considerable amount of lab work. Nevertheless, this theory is attractive because it does provide a possible path to RNA or DNA.

The Wächtershäuser model for the origin of protometabolism also relies on iron and sulfur chemistry. This scientist (who is also a patent lawyer) does away altogether with the notion of primordial broth. For him, Earth did not need a helping hand from space, lightning, or a reducing atmosphere; all necessary ingredients were present from the beginning in volcanoes and hydrothermal vents. He considers that only CO_2, CO, H_2S (present today in volcanic emissions and hydrothermal vents), and FeS (common in Earth's crust) were needed to get protometabolism started.

This is how it would have worked. First, as in the de Duve model, a source of energy and electrons (to achieve reduction reactions) is necessary. This source could have been the reaction between iron sulfide and hydrogen sulfide:

$$FeS + H_2S \rightarrow FeS_2 + 2\,e^- + 2\,H^+,$$

where FeS_2 is pyrite. In addition to releasing electrons for reduction, this reaction also releases energy. The mineral pyrite formed in this reaction has interesting properties, in that it can strongly bind to all sorts of electrically charged molecules. This, according to Wächtershäuser, would have allowed organic molecules to line up in close proximity on the mineral's surface and undergo further chemical reactions, a protometabolism of sorts. For example, the universal metabolite pyruvic acid (see chapter 3) can actually be formed in a sequence of three reactions:

$$CO_2 + 2FeS + 2H_2S \rightarrow CH_3{-}SH + 2FeS_2 + 2O$$
$$CH_3{-}SH + CO \rightarrow CH_3{-}CO{-}SH$$
$$CH_3{-}CO{-}SH + CO_2 + FeS \rightarrow CH_3{-}CO{-}COOH + FeS_2,$$

where CH_3—CO—COOH is pyruvic acid. Other protometabolic reactions would have occurred in a similar fashion, including the synthesis of amino acids and nucleotides, necessary to make proteins and RNA, respectively. In this scenario, there is no need for thioesters (as in de Duve's model). Nor is there need for organic material to be delivered from outer space or from a reducing atmosphere. On the other hand, this model provides no clue regarding the functions of putative protoenzymes and first nucleic acids.

Finally, the American biochemist Sydney Fox proposed that protein enzymes could have been produced on prebiotic Earth in the absence of genetic information. For this, he first demonstrated that amino acids analogous to those found in a Miller-type of experiment could also be synthesized by simply heating aqueous solutions of formaldehyde and ammonia. These two compounds can be formed under a variety of plausible prebiotic scenarios. Next, he demonstrated that heating dry mixtures of amino acids leads to the production of polymers containing up to several hundred chemically linked amino acids, thus mimicking modern proteins. Fox used dry heat in these experiments for the following reason: the linking of amino acids to produce protein polymers cannot happen in water because this reaction is accompanied by the elimination of a water molecule. Thus in an aqueous environment, the water molecules present in the solution favor the reverse reaction—that is, the destruction of the bonds linking amino acids.

Fox reasoned that some of the primeval aqueous broth with its amino acids could have been splashed by the wind or waves onto hot rocks, where the water evaporated and the amino acids started polymerizing. This effect would have created a large variety of randomly formed protein-like molecules. Some of these molecules could have been endowed with enzyme activity and could have started a kind of protometabolism, as assumed by de Duve in the case of thioester enzymes. The question then is, do protein-like polymers "à la Fox" made in the laboratory possess any kind of enzyme activity? The answer is yes, they do, albeit weakly. Furthermore, Fox's proteinoids, as they are called, can spontaneously form microscopic spheres in which molecules that will be acted upon by proteinoids can be trapped. Thus these microspheres could be seen as primitive cells, possibly performing some protometabolic functions.

The main line of criticism aimed at these results and interpretations is that such microspheres, even if they were formed and persisted (they are quite unstable and fragile under laboratory conditions), could not have evolved because they contain no genetic information. Of course, one could always retort that proteinoid microspheres could have become capable of synthesizing random nucleic acids that were themselves able to evolve. Fox died in 1998 and, to my knowledge, work on proteinoids has stopped.

GENETIC INFORMATION FIRST: THE RNA WORLD

Not everyone is buying the iron-sulfur world hypothesis and its haphazard synthesis of protein catalysts in the total absence of genetic information. This skepticism also applies to Sydney Fox's theory. Many scientists (possibly a majority) prefer to think that genetic information came first and that proteins followed suit, once some kind of transcription/translation mechanism had evolved. For this to have happened, it must be assumed that informational macromolecules such as DNA or RNA must have been somehow synthesized in the absence of protein enzymes. In the late 1960s, three scientists, Carl Woese of the United States, Leslie Orgel of England (but working in San Diego), and Nobel laureate Francis Crick of England (also working in San Diego) independently proposed that RNA, not DNA, could have been the first genetic material.

There are four reasons for this. First, many viruses possess an RNA genome. Second, the synthesis of DNA building blocks, deoxyribonucleotides (consisting of nitrogenous bases linked to the sugar deoxyribose and to phosphate groups), proceeds via ribonucleotide (the building blocks of RNA) intermediates. Third, DNA replication in cells is "primed" by short stretches of RNA. And fourth, to be expressed, DNA genes must first be transcribed into RNA molecules, which are then decoded by transfer RNAs on a ribosomal matrix composed of 50 percent RNA. These observations suggested that RNA may be more ancestral than DNA and may have constituted the very first genes.

This hypothesis raises several questions. The first one pertains to the likelihood of synthesizing the building blocks of RNA, the ribonucleotides, made of the bases A, U, G, and C, the sugar ribose, and three phosphate groups. The second question deals with linking these blocks together to form RNA chains, and the third problem is how to make these RNA chains replicate (multiply) to keep them going so that they evolve into genuine living cells. Today, all three mechanisms involve enzymes. In an RNA world, there were no enzymes, and it must thus be assumed that ribonucleotide synthesis, polymerization, and RNA replication either occurred spontaneously or were catalyzed by nonenzyme catalysts.

We have seen that the four bases could have been formed from hydrogen cyanide or other carbon- and nitrogen-containing molecules under reducing conditions. Alternatively, these bases could have been brought to Earth by meteorites and comets (although cytosine has never been found in meteorites). The sugar ribose could have been formed from the polymerization of formaldehyde, also presumably abundant in the prebiotic broth, as it is in space. The problem here is that the polymerization of formaldehyde produces many other types of sugars besides ribose, and these other sugars also have a propensity to react with the four bases. Some authors have recently proposed that in the beginning, nucleic acids other than RNA may have been more prevalent. These include molecules formed with sugars containing six carbon atoms instead of the five found in the ribose ring, or a five-carbon-containing sugar (called threose, hence the abbreviation TNA for the nucleic acids containing it) with a structure slightly different from that of ribose. Thus it is not impossible that originally there was a whole family of nucleic acids whose members contained different sugars, with RNA finally taking over for unknown reasons. According to some, phosphorus could have originated from organic phosphates found in meteorites.

So far, linking the four RNA bases to sugars in the laboratory, under plausible prebiotic conditions, has been successful only with A and G. Such reactions with U and C have not been observed. This is definitely a nagging problem, because it is hard to visualize RNA synthesis without the existence of ribo-U and ribo-C. Can these compounds be formed in the presence of special, and as yet undiscovered, catalysts? Were TNAs involved? We simply do not know. On the other hand, adding phosphate groups to preformed combinations of bases and ribose has been achieved with all four bases. The problem here is that many of the base-sugar-phosphate combinations (called nucleotides) have structures that would not allow polymerization into RNA molecules. Clearly, much remains to be done to elucidate the mechanisms by which correct structures where formed in prevalent amounts. One possibility is that mineral catalysts favored the formation of correct molecular structures, but such catalysts have not yet been discovered. Alternatively, it might be imagined that the correct molecules were formed with the help of protoenzymes and pyrophosphate previously made in an iron-sulfur world.

Also, sugars found in modern nucleic acids pose a handedness problem, just as seen with amino acids. The sugars present in RNA and DNA are right-handed. However, prebiotic sugars must have been made in the right- and left-handed configurations in equal amounts. We do not know how right-handed sugars were eventually selected for the synthesis of nucleic acids. Nevertheless, assuming that pools of nucleotides with the correct structure came into existence, it is now necessary to link these nucleotides together to form RNA chains. Interestingly, this has been achieved in the laboratory by simply incubating nucleotides in the presence of minerals such as lead salts, uranium salts, zinc salts, or even clay. RNA chains consisting of up to fifty bases linked together were synthesized on a clay (montmorillonite) substrate. Quite certainly, these findings make plausible the synthesis of RNA chains under prebiotic conditions.

But then there remains the problem of replicating these RNA molecules. Without replication, the genetic information present in an RNA molecule will be lost as soon as this molecule degrades. Furthermore, without replication and errors in the replication process, RNA molecules once formed cannot change— they cannot evolve. Perhaps surprisingly, RNA replication in the absence of enzymes may not have been as daunting a challenge as was once thought. Clues to the prebiotic copying of RNA molecules were provided by the discovery of ribozymes, RNA enzymes.

Until 1983, that all biological catalysts were protein enzymes was a firmly entrenched notion. That year, Thomas Cech and Sidney Altman independently discovered that this dogma was wrong. They both received the Nobel Prize for this discovery. Their findings unexpectedly showed that RNA too possesses catalytic, enzymatic activity, hence the name ribozyme (a combination of the words *ribonucleic acid* and *enzyme*). What kind of enzymatic activities are displayed by RNA molecules? First, there are conditions under which living cells "cut" some of their RNA molecules by cleaving their ribose-phosphate backbone at predetermined positions (figure 4.7). This cutting is done by the RNA itself, with no help whatsoever from protein enzymes. Next, it was discovered that ribozymes could actually copy themselves and thus replicate, again, in the total absence of protein enzymes! It must be cautioned, however, that only short RNA molecules, a few dozen nucleotides in length, can be replicated that way in the laboratory. And finally, it was recently discovered that ribosomal RNA (present in the ribosomes, where protein synthesis is taking place in cells) is responsible for the polymerization of amino acids into proteins. In other words, it is not a ribosomal protein enzyme that hooks up amino acids together to make protein chains; ribosomal RNA itself does the job. Some ribozymes even

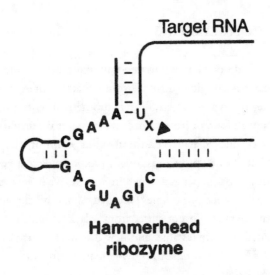

FIGURE 4.7 A ribozyme cutting another RNA molecule. A natural ribozyme, called the hammerhead ribozyme, binds to a target RNA through base pair formation (*short vertical and horizontal stripes*). The sugar-phosphate backbone of the target RNA is cut to the right of the base designated X (which can be any base). (Adapted from Breaker, R. R. 2000. Making Catalytic DNAs. *Science* 290:2095-2096.)

have the ability to splice short RNA molecules together to produce longer ones. Finally, RNA chains consisting of as few as twenty nucleotides have been shown to possess ribozyme activity. This number is well within the ability of clay to catalyze the synthesis of RNA chains.

The discoveries that certain RNA molecules can cut themselves and other RNA molecules, that they can join RNA molecules together, that some ribozymes can replicate, and that certain RNA molecules can catalyze the formation of proteins have given credence to a putative RNA world. In this prebiotic world, primitive RNA genomes could replicate, become processed by cutting and splicing, and help make proteins that later could become genuine enzymes, taking over some of the ribozyme properties and creating an integrated metabolic circuitry. Before this could happen, RNA genomes had to evolve through replication and replication errors in order to generate genetic diversity, much as natural selection does in the world today. We will see how this may have happened in the next chapter.

CONCLUSIONS

We do not know how the building blocks of life appeared on Earth. They may have originated from organic material present in interstellar clouds, from meteorites and comets, from hydrothermal vents, from a reducing atmosphere, from all four, or from sources we have not yet imagined. Whether proteins antedated nucleic acids in an iron-sulfur world or whether an RNA world gave birth to the first proteins is equally unknown. Perhaps even the iron-sulfur and RNA worlds cooperated to get life started. We have several scenarios but we do not know which one or ones prevailed. What this chapter has shown, however, is that scientists do not suffer from a lack of imagination. The hypotheses presented here are testable, and new discoveries, such as the catalytic properties of RNA, will continue to provide material for further research and hypothesis formulation.

In 1998, the British Broadcasting Corporation released a video entitled "The Origin of Life: Researching the Possibilities." I highly recommend it to readers, not only for its informational content and its general quality but also for its humorous look at the problem. The program features among others, Stanley Miller (in front of his famous "gas zapper") and Günter Wächtershäuser (handling a block of his cherished pyrite). Consciously or unconsciously, some of the scientists appear as enthusiastic nerds[3] (one of them, unable to find a chalkboard to explain something to the interviewer, casually and very naturally uses his office door to write on), vindictive prima donnas, or self-assured wise men. This video is also a lesson in the sociology of science.

CHAPTER 5

Life on Its Way

"What impresses me most," he continued, "is that everything comes from one single cell. Several million years ago a little seed appeared which split in two, and as time passed, this little seed changed into elephants and apple trees, raspberries and orangutans. Do you follow me, Hans Thomas?"

—JOSTEIN GAARDER, *The Solitaire Mystery* (1996)

cience sometimes works in strange ways. One might think that basic principles, such as the origin of the building blocks of life, should be firmly established before starting a discussion of the next step. But this is not necessary if one simply assumes that these building blocks *did* appear somehow or other. Otherwise, we would not be here to think about the origins of life. I will assume in this chapter that the chemistry of the prebiotic broth synthesized RNA molecules capable of replication. Whether protoenzymes made in an iron-thioester world were involved in the replication mechanism is not that relevant here, because this chapter deals with the evolution of the RNA world and its transformation into a world where cells, no longer isolated molecules, dominated. The focal point of this chapter is the problem of the evolution of the primitive genetic material into the sophisticated information storage and transfer mechanisms that now operate in living cells. Not surprisingly, some of the ideas developed by scientists interested in the origins of life derive directly from studies done with extant biological systems. One such system of particular interest is constituted by bacteriophages.

LESSONS FROM BACTERIOPHAGES

Simply put, bacteriophages are viruses that infect and kill bacteria. Most bacteriophages have a simple structure, consisting of nucleic acid packaged in a coat made of protein. Many bacteriophages have a genome made of RNA that replicates many times in the infected host before its death occurs. The enzyme that makes possible the multiplication of a bacteriophage RNA genome is called an RNA replicase, and the gene that codes for this enzyme is present in the bacteriophage genome itself. To replicate this genome, the RNA replicase "reads" it—that is, it uses it as a template to link the relevant nucleotides together and form a new viral RNA chain. This process is repeated many times to produce a large number of copies of the invading bacteriophage RNA. It has been known for quite some time that RNA replicase is not a very precise enzyme; it makes mistakes at a fairly high rate by incorporating nucleotides at wrong positions in the newly formed RNA molecules. RNA replicase thus frequently produces mutant copies of the original invading bacteriophage RNA.

In 1970, Sol Spiegelman of Columbia University published the results of an evolution experiment conducted in the test tube. He mixed RNA isolated and purified from bacteriophage Qβ (this RNA has a length of 4500 nucleotides), purified Qβ replicase, and all four nucleotides [adenosine-, guanosine-, cytidine-, and uridine triphosphates (ATP, GTP, CTP, and UTP)] to serve as building blocks for newly synthesized Qβ RNA molecules. It was known at the time that this simple test tube system worked very well and that copious amounts of Qβ RNA could be made this way. What was new, however, was the fact that Spiegelman did not wait until his test tube was full of newly made Qβ RNA; he did a number of serial transfers in which a drop from the first test tube was added after a short time to a second test tube containing the replicase and nucleotides but no new RNA. The process was repeated a third time, and so on (figure 5.1).

After a number of transfers, Spiegelman noticed that the RNA produced was no longer the 4500-base-long initial Qβ RNA; rather, the new RNA population consisted of a set of RNA molecules of approximately 500 bases, which he called variants. What had happened? Remembering that replicases are somewhat sloppy enzymes, these variants were produced when the replicase fell off the RNA template before fully completing its replication. These short RNAs replicated much faster than the full-length Qβ RNA simply because, as they were much shorter, they successfully competed with the longer RNA molecules for

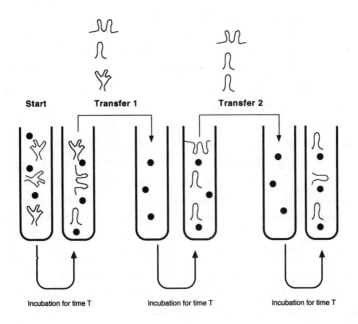

FIGURE 5.1 Serial transfer experiment with the bacteriophage Qβ system. The first test tube contains the four nucleotides ATP, GTP, CTP, and UTP, Qβ replicase (*dots*), and Qβ RNA molecules (*folded lines*). After a certain time, a small sample from this tube is transferred to a second tube also containing the four nucleotides and Qβ replicase but no additional Qβ RNA. Again, a sample of the second tube is transferred to a third tube, and so forth. At the end of the experiment, the newly synthesized Qβ RNA is analyzed. It can be seen in this example that at the beginning of the experiment, three-digit RNA molecules are present. After the second transfer, a shorter, single-digit population is observed. In a real experiment, many transfers are necessary to achieve selection of a particular population of RNA molecules. (Adapted from Smith, J. M., and E. Szathmáry. 1995. *The Major Transitions in Evolution.* Oxford, England: Oxford University Press.)

binding to the replicase. After enough transfers, there was nothing left of the original Qβ RNA; the test tube was now full of short variants. This truly was evolution in the test tube: the short variants outcompeted the long RNA because, being short, they possessed better fitness in the replication process.

These experiments were taken a step further in the laboratory of Manfred Eigen of Germany, a Nobel laureate in chemistry. There researchers mixed Qβ replicase with the ATP, GTP, CTP, and UTP nucleotides but added *no* Qβ RNA template at all. To what must have been everyone's surprise, RNA was synthesized! In other words, the replicase was able to make RNA without a template to copy. This also meant that a protein (the replicase) was able to produce a genetically informational macromolecule (RNA) without preexisting genetic information! Could this have happened in the prebiotic broth? We do not know, but it remains an intriguing possibility.

What is more, the RNA molecules were not of just any kind. They formed a family of short molecules composed of 150 to 250 nucleotides (therefore very much shorter than genuine Qβ RNA), called minivariants. Surprisingly, a 220-nucleotide-long species, one of the minivariants, is also found in nature in bacteria infected by Qβ. How is one to interpret all this information? First, the fact that a natural 220-nucleotide-long RNA was produced in the test tube indicated that this experiment was not completely off the wall. Next, this experiment also showed that a whole family of RNA molecules was made under these conditions, not just one single type of RNA. This is important because considerable genetic variation was generated in this system, thanks to the ability of the replicase to make mistakes in the replication process.

Taken together, the results of Spiegelman and Eigen show how a simple system can produce *variation*, and that selective pressure (in the form of serial transfers) can *select* for a particular family of variants. In fact, something similar may have happened in a pure RNA world without protein enzymes. Indeed, it is known that RNA replication catalyzed by ribozymes is imperfect, and therefore the RNA world would also have been able to create RNA variants. In other words, the existence of RNA variants in the prebiotic world made evolution by natural selection possible, with the most fit RNA molecules outcompeting the less fit. In the beginning of a self-replicating RNA world, better fitness simply meant more efficient replication, and hence better ability to compete for "food," the nucleotide building blocks of RNA.

QUASISPECIES AND HYPERCYCLES

Let us now consider a prebiotic world where a set of RNA molecules capable of replication (with or without protoenzymes) coexist in a pond with a given supply of nucleotides necessary for their replication. What is the fate of this population of RNA molecules? To tackle this problem, it is useful to think in terms of a quasispecies—a population of RNA molecules composed of individuals all possessing slightly different base sequences produced by replication errors. Or, to put it differently, a quasispecies is an ensemble of minivariants. These minivariants derive from what is called a master sequence, which is a single sequence representing the highest probability of finding a particular base at a particular position in that sequence. For example, the five short RNA molecules in the following list:

UCGUCCA
AAUUACG
ACAAAUG
ACGUGCG
ACGCACG

derive from the master sequence ACGUACG.

In this example, it can be seen that the first minivariant has a U in first position, whereas the other four minivariants have an A (in bold) in that position. Thus an A has the highest probability of figuring in that first position and is found there in the master sequence. The second minivariant has an A in second position and a U in third position. We can see that the other variants have a C in second position, and the third position is occupied by a G in three out of five variants. Hence, the most probable base in second position is a C (in bold), and in third position it is a G (also in bold), and so on. Thus the most probable positions of bases are found in the master sequence.

Let us now assume that the master sequence has the highest fitness and replicates faster than the variants surrounding it. Knowing that RNA replication in the prebiotic world was imperfect, it is legitimate to ask what a particular replication error rate will do to the existence of a quasispecies and its master sequence. It can be demonstrated that

$$N < \ln s/(1 - q),$$

where N is the length (in number of bases) of the RNA, $\ln s$ is the natural logarithm of the fitness of the master sequence, and q is the fidelity of replication of each base. The value of q varies between 0 and 1; if $q = 0$, there is a 100 percent error rate at the level of that base, and if $q = 1$, that base is faithfully replicated each time. Therefore the error rate per base per round of replication is $(1 - q)$. This equation is called a hyperbolic function. The graph representing that function is shown in figure 5.2. In it the y-axis represents RNA length and $(1 - q)$ on the x-axis is error rate. The region of the graph where a quasispecies can exist stably is represented by the hatched area. Outside this area, demarcated by the hyperbola, a quasispecies will disappear because the replication

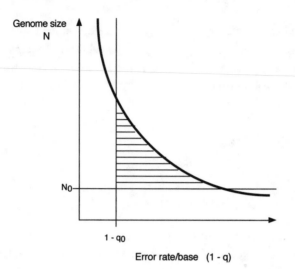

FIGURE 5.2 The replication error threshold. The hyperbola (*thick line*) represents the threshold, and the *hatched area* represents the region where quasispecies characterized by a genome size N and a certain error rate can survive. N_0 is the lower limit for sufficient encoded information (50 to 100 nucleotides). The number $(1 - q_0)$ is a limit below which replication energy and time are prohibitive for a system. (Adapted from Smith, J. M., and E. Szathmáry. 1995. *The Major Transitions in Evolution*. Oxford, England: Oxford University Press.)

error rate is too high. The hyperbola thus defines an error threshold beyond which quasispecies stop constituting "clouds" of variants surrounding a given master sequence. As can be seen from the graph, smaller RNA molecules tolerate higher error rates than large RNA molecules. The validity of this equation has been verified in living cells; for example, the error rate for bacteriophage Qβ is 5×10^{-4} for a length of 4500 bases, a value that puts it below the error threshold. This is why bacteriophage Qβ still exists today.

These theoretical developments clearly show two things: first, considering that the error threshold in the RNA world must have been very high, it can be calculated that the maximum size of the RNA molecules that could have been maintained under prebiotic conditions was about 100 nucleotides. This is a short length but still above that of modern transfer RNA molecules (seventy to eighty nucleotides). Second, RNA quasispecies displaying too high a replication error rate were doomed to extinction. This does not mean that all the RNA molecules present in this quasispecies necessarily disappeared; it means that sequences became randomized and no longer derived from a single master sequence. However, some of the RNA molecules produced under conditions exceeding the error threshold could have themselves established new, better-adapted quasispecies that replaced the extinct ones. In other words, the RNA world was able to evolve.

The next problem then is to understand how and why short prebiotic RNA molecules did not simply evolve into one gigantic quasispecies particularly fit to replicate but perhaps unable to perform any other function, such as coding for proteins. In this scenario, the "living" world today would still consist of that one successful RNA quasispecies. Eigen came up with the notion of hypercycles to solve this conundrum. A hypercycle is a circular feedback system of replicators (such as RNA molecules) in which each replicator depends on the success of the other replicators. A simple hypercycle is shown in figure 5.3. Basically, the stability of a hypercycle depends on the degree of cooperation contributed by each of its members.[1]

This whole concept is further complicated by the fact that each member of a hypercycle is a quasispecies. However, this complication has an advantage: hypercycles, too, can evolve. Let us imagine that one of the members of the hypercycle evolves into a quasispecies that not only is able to replicate itself faster but also helps the other members of the hypercycle to replicate faster. This is possible if one of the quasispecies mutates into an efficient replicating ribozyme that can replicate the other members of the hypercycle as efficiently as itself. In that case, this hypercycle will outcompete the other hypercycles. If, on the contrary, one of the quasispecies mutates into a "selfish" replicating ribozyme that replicates itself efficiently but somehow inhibits the replication of the other members, that hypercycle will disappear. Successful hypercycles would then have been those with the most versatility, in particular those that would have started coding for protein enzymes capable of improving the fidelity of RNA replication and the ability to make their own building blocks rather than simply relying on the prebiotic broth.

There is a problem with hypercycles, however. As they are made of free-floating RNA molecules presumably dispersed in water, it is hard to see how stable hypercycles could have existed in the absence of membranes containing them. For example, a rock falling into the pond would have disturbed any unbounded hypercycle, however successful it may have been before this accident. There was thus a need for hypercycle encapsulation. How did this happen? As we saw in

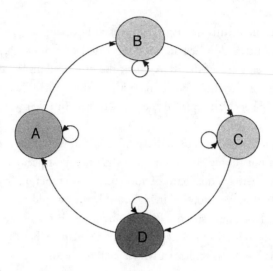

FIGURE 5.3 A simple hypercycle. The quasispecies A, B, C, and D are replicating quasispecies (as indicated by the *small circular arrows*) that may also possess other ribozyme activity. In this scheme, each of the four quasispecies depends on the replicative success of the preceding one, as indicated by the *large arrows*.

chapter 3, modern cell membranes are based on phospholipid bilayers. These are not produced in sparked gases and it is not known whether hydrothermal vents synthesize them. On the other hand, amphiphilic molecules, such as octanoic acid (with eight carbon atoms) and nonanoic acid (with nine carbon atoms) are found in meteorites and could have been used to encapsulate hypercycles. Biochemists have recently demonstrated that one octanoic acid, a carboxylic acid with the formula $CH_3(CH_2)_6COOH$, potentially present on prebiotic Earth, can spontaneously form vesicles (technically called liposomes) when dispersed in water. Liposomes can trap nucleic acids (RNA and DNA) if they form in a solution containing them; alternatively, preformed liposomes pulsed in water with an electric discharge in the presence of nucleic acids can also encapsulate them (figure 5.4). Once again, lightning, as in Miller's experiment but this time striking a pond containing both liposomes and RNA, could have led to the formation of "protocells," entities made of one or several hypercycles contained within a membrane.

Interestingly, liposomes made of amphiphilic molecules containing eight or nine carbon atoms are permeable to substances dissolved in water. Therefore

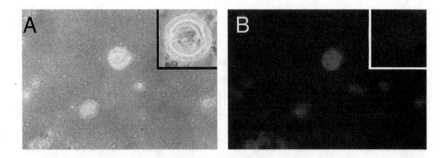

FIGURE 5.4 Liposomes having trapped nucleic acids after submission to electric discharge. **A:** Micrograph taken in visible light. **B:** Micrograph taken in UV light. Nucleic acid molecules were tagged with a dye that fluoresces red under UV illumination. **Inset:** A control liposome that was not electrically shocked is not fluorescent and has not incorporated any nucleic acid molecules. (Adapted from Lurquin, P. F., and K. Athanasiou. 2000. Electric field-mediated DNA encapsulation into large liposomes. *Biochemical and Biophysical Research Communications* 267:838-841.)

prebiotic liposomes containing RNA would have been able to import the building blocks necessary for RNA replication. What is more, laboratory experiments have shown that vesicles made of octanoic acid can grow spontaneously by incorporating free octanoic acid present in the medium. This could have led to some primitive type of liposome division in the prebiotic world. Modern cell membranes are made of phospholipids containing sixteen to eighteen carbon atoms. Liposomes made from such phospholipids are highly impermeable to water-soluble compounds. Living cells also contain proteins embedded in their phospholipid bilayers, and these are responsible for the import and export of nutrients and waste products. One must then assume that short amphiphilic molecules were replaced by long ones only after the appearance of proteins.

ORIGIN OF THE GENETIC CODE: FROM THE RNA WORLD TO PROTEINS

As we saw in chapter 3, in modern cells the genetic information stored in DNA is first transcribed into messenger RNA molecules. This step was unnecessary in the RNA world because genes were made of RNA, not DNA. As we also saw, mRNA molecules are next decoded by transfer RNA molecules to which an amino acid is attached. This process, called translation, results in the synthesis of proteins. In translation, the positioning of tRNAs along mRNA molecules is determined by sequences of three bases, called codons, that are read by the tRNA anticodons which also contain three bases. Now, two important questions arise. First, how did the RNA world develop the ability to attach amino acids to tRNAs, and second, what is the origin of the genetic code that specifies which codon corresponds to which amino acid? In modern cells, amino acids are attached to tRNAs through the action of protein enzymes called aminoacyl-tRNA ligases (figure 5.5). This was not possible in the RNA world, where proteins did not exist.

Manfred Eigen's team has done much work in this area. Much like DNA, RNA can form double helical structures through specific interactions between A and U and G and C. Thus primitive tRNAs and primitive mRNAs could have interacted by double helix formation, as they do in modern cells. For example, a proto-tRNA with a CCG anticodon could form a very short double helix with a proto-mRNA molecule containing a GGC codon and lock the amino acid it carries into place. Therefore the decoding of the genetic message in mRNAs by tRNAs is simply an intrinsic property of RNA. It is less clear how tRNA molecules "learned" how to attach specific amino acids. Some modern tRNAs can bind amino acids through the formation of pockets in their structure. Indeed, RNA molecules should not be seen as flat strings of nucleotides; they can assume complicated three-dimensional configurations. Proto-tRNAs may have been able to trap amino acids in the same way. Furthermore, a new type of ribozyme activity has been discovered recently: some short RNA molecules synthesized in the test tube can chemically bind amino acids without the help of protein enzymes. This activity could have been common in the RNA world.

Following up on his idea of quasispecies, Eigen asked whether tRNAs could once have existed as a set of molecules grouped around a master sequence. To put it differently, Eigen wondered whether it was possible to trace the lineage of modern tRNAs to an ancestral master sequence that existed billions of years ago.[2] If so, this would add great credibility to the concept of an RNA world. To answer this question, Eigen and coworkers compared the base sequences of 200 modern tRNA molecules, isolated from humans, animals, plants, fungi, and bacteria, and applied to these sequences the principle of the phylogenetic tree that we saw in chapter 3. And indeed, a master sequence was found! Figure 5.6 shows the sequence of "the mother of all tRNAs." The fact that this sequence folds into the classical cloverleaf configuration of modern tRNAs adds great weight to the notion that tRNAs were (and still are) a quasispecies. The first

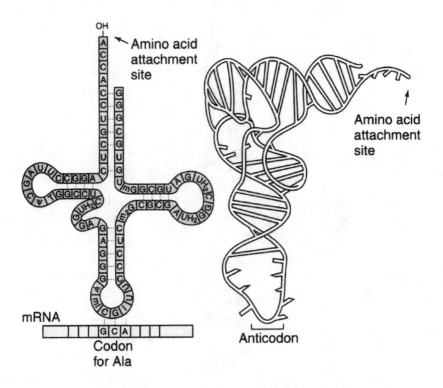

FIGURE 5.5 Amino acid attachment to a tRNA molecule. **Left:** Two-dimensional cloverleaf structure of a tRNA molecule. The amino acid binds to the end of the tRNA designated —OH. **Right:** Three-dimensional structure of the same tRNA. On the left, the anticodon of the tRNA is seen interacting with a GCA codon (coding for alanine) in the mRNA. I, mI, Ψ, UH$_2$, and mG are special bases found in tRNAs only. (Adapted from Hartwell, L. H., L. Hood, M. L. Goldberg, A. E. Reynolds, L. M. Silver, and R. C. Veres. 2000. *Genetics.* New York: McGraw-Hill.)

time I saw this molecule, I could not help experiencing a slight shiver and a temporary mental blank. I was looking at a molecular fossil that existed at the dawn of life!

Next, it became necessary to tackle the origin of the genetic code, which, we know, is composed of sixty-four different codons determining twenty amino acids. It is highly unlikely that this complicated code appeared all at once. By taking a close look at the anticodon portion of the tRNA master sequence (the portion that reads the codon in the mRNA), Eigen and coworkers realized that this master anticodon was able to decode codons of the RNY type, where R is A or G, N is any base, and Y is C or U. In the prebiotic world, in the absence

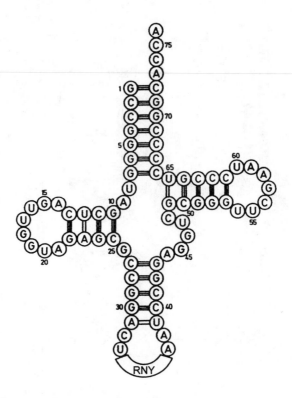

FIGURE 5.6 The ancestral tRNA according to Eigen and coworkers. (Adapted from Eigen, M., and R. Winkler-Oswatitsch. 1981. Transfer-RNA: The early adaptor. *Naturwissenschaften* 68:217-228.)

of stabilizing proteins, it would have been important for proto-tRNAs and proto-mRNAs to interact as strongly as possible during the decoding process. Knowing that the interactions between G and C in codon-anticodon binding are stronger than A-to-U interactions, the RNY codons must have been the following four:

GGC, which codes for glycine
GCC, which codes for alanine
GAC, which codes for aspartic acid
GUC, which codes for valine

In other words, the first genetic code would have been able to determine only four amino acids. Does this make sense? It certainly does when we consider that these are the four amino acids that are the most prevalent in a Miller-type experiment! This of course does not prove that Earth's primeval atmosphere was strongly reducing (as required to make amino acids in the atmosphere) or that Eigen's interpretation is correct. Yet, if this is a coincidence, it is a strange one.

Thus the original genetic code could have been unidimensional, as shown in figure 5.7. How did it become three-dimensional? The mechanisms are not known, but we can assume that it first became two-dimensional by addition of an A in the first position and a C or a U in third position. Three-dimensionality was finally achieved by addition of more bases (see figure 5.7).

We have now reached a point where proto-tRNAs in the RNA world could bind amino acids and also decode the codons found in proto-mRNAs. This thus allowed the positioning of amino acids next to one another in very close proximity. To form a protein, one last step needed to be performed—the linking of adjacent amino acids via chemical bonds. As we saw earlier, modern ribosomal RNA is capable of linking together amino acids in a growing protein chain through ribozyme activity. This could also have happened in the RNA world through the action of proto-ribosomal RNA molecules. When all these activities became integrated—that is, as soon as hypercycles (consisting of membrane-bound proto-tRNAs, proto-ribosomal RNAs, and proto-mRNAs) became able to synthesize proteins—the first protocells were born. This was a giant leap in the direction of life as we know it.

Putting aside the complexity of the RNA world and its evolution, the following is a summary and tentative sequence of events that took place in it.

Short proto-tRNAs and proto-mRNAs, formed randomly by the linking of nucleotides, exist as replicating quasispecies. They replicate via ribozyme activity and cooperate via formation of hypercycles.

Proto-tRNAs capture the four most prevalent amino acids, also via ribozyme activity. The proto-tRNAs interact with proto-mRNAs via GNC codons.

Proto-tRNA-proto-mRNA interactions through codon-anticodon binding bring amino acids in close contact.

Ribozyme activity carried by proto-ribosomal RNA (also a quasispecies) links the amino acids together to make simple proteins with primitive enzymatic activity.

Ribozyme activity splices together short proto-mRNAs to make longer proto-mRNAs with enhanced coding ability.

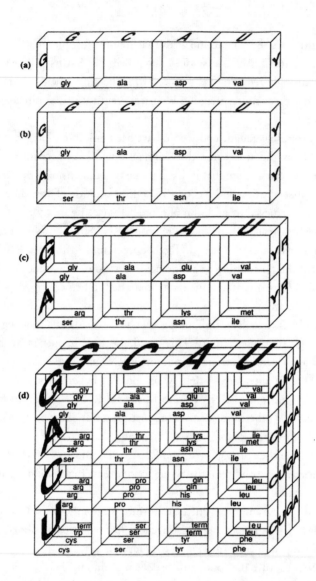

FIGURE 5.7 Evolution of the genetic code. The primitive unidimensional code is shown in (a) evolved into another unidimensional code and then in (b) with the addition of a second base (Y) that represents U or C. A two-dimensional code (c) then appeared by addition of an R base, representing either A or G. This code then gave birth to the present, three-dimensional code shown in (d). Amino acid abbreviations are as in figure 3.2. (Adapted from Eigen, M., and R. Winkler-Oswatitsch. 1981. Transfer-RNA: An early gene? *Naturwissenschaften* 68:282-292.)

At this point, or conceivably earlier, the hypercycles, consisting of proto-tRNAs, proto-mRNAs and proto-ribosomal RNAs, become encapsulated by membranes. Once this happens, evolution by natural selection gains full force, as the fittest protocells start proliferating faster than the less fit.

Finally, the genetic code starts evolving as proto-enzymes catalyze the synthesis of amino acids other than the four initial ones. As a result, protein (enzyme) variability and versatility increase. At a certain point, RNA replicases, able to replicate RNA molecules with much more fidelity than ribozymes, are born. This higher fidelity allows the propagation of longer mRNA molecules, and diversity increases further. Life, in the form of RNA-containing protocells, is definitely on its way.

As a final note, it should be stressed that the iron-thioester world and the RNA world are not two completely incompatible models. In the iron-thioester world, protoenzymes formed by spontaneous thioester condensation would have helped the various steps just described. These protoenzymes would then have been taken over by proteins made through genuine, RNA-encoded proto-translation.

PROTOCELLS AND THE EMERGENCE OF THE DNA WORLD

What would the very first protocells have looked like? First, since DNA did not exist yet, they must have been ribo-organisms with a genome made of RNA. Next, their metabolism must have been very simple. They may have been chemoautotrophs (see chapter 3), relying entirely on mineral nutrients found naturally in the environment. One possible scenario for their overall metabolism was that they reduced atmospheric carbon dioxide in the presence of hydrogen and oxidized iron sulfide into pyrite with the help of hydrogen sulfide:

$$4\ CO_2 + 7\ H_2 \rightarrow (CH_2COOH)_2 + 4\ H_2O,$$

where $(CH_2COOH)_2$ is oxalic acid, a compound that can participate in other oxidation-reduction reactions.

Hydrogen was provided by the formation of pyrite:

$$FeS + H_2S \rightarrow FeS_2 + H_2,$$

where FeS is iron sulfide and FeS$_2$ is pyrite.

The reduction of carbon dioxide into oxalic acid releases energy that could have been stored in ATP or pyrophosphate, or both, and used to drive other cellular reactions. Protocells evolved through mutations introduced into their genome by ribozyme replication errors and, later on, by RNA replicase-induced mistakes. It is also possible that gene duplication, that is, the formation of two copies of the same gene inside one cell, accompanied the appearance of protocells. We know that gene duplication is one of the main engines of evolution. Indeed, when an essential gene is duplicated, as long as one copy of this gene remains intact, the other copy can mutate without causing harm to the cell. As mutations accumulate in the second copy of the gene, this gene diverges more and more from its companion.[3] At a certain point, the mutant duplicated gene can start coding for a protein very different from that coded for by the original gene. Entirely new functions can appear under these circumstances. It is very possible that protocells rapidly diverged by RNA gene duplication.

At a certain point, RNA genomes were converted into DNA genomes and the DNA world we know today was created. DNA genomes, since they took over almost all life-forms (except some viruses), must have had evolutionary advantages. What were they? The chemistry of DNA shows that it has at least three advantages over RNA as far as genomes are concerned. First, DNA is more stable than RNA in the cell's environment, because thymine and deoxyribose are more chemically stable than their RNA counterparts, uracil and ribose. Natural selection would, of course, have favored cells whose genomes had become more stable, more long lasting. Second, it is known that ribozyme activity is the result of the presence of an extra OH group in ribose, the sugar found in RNA. Since DNA contains deoxyribose and thus does not have that extra OH group, it would have been unable to engage in the many ribozyme activities we have studied. This was a definite advantage once long genes were created. Indeed, ribozyme activity, important as it was at the very beginning of life, always runs the risk of cutting large RNA molecules into smaller pieces. This would have destroyed long RNA genes. This problem was thus solved with the appearance of DNA genes. Finally, the creation of DNA completely separated gene replication and translation. In the RNA world, ribozyme activity accomplished both at the same time, meaning that a serious error at one level would have created serious errors at all levels. This is avoided in the DNA world.

But then, how is RNA converted into DNA? This must have happened in the distant past through the creation of new genes and the mutation of preexisting ones. Remember that in modern cells, the building blocks of DNA are first synthesized as RNA building blocks. Thus deoxyribose is made from ribose through the action of an enzyme that eliminates the extra OH group from ribose. This enzyme is called a ribonucleotide reductase. Furthermore, the DNA base thymine is produced by the addition of a methyl (CH_3) group to the RNA base uracil through the action of a methylase enzyme.

Then, the deoxyribonucleotides dATP, dGTP, dCTP, and dTTP (the last being deoxythymidine triphosphate) must have been used to synthesize DNA by copying RNA gene templates. This is accomplished today by enzymes called reverse transcriptases. The virus that causes acquired immunodeficiency syndrome (AIDS), for example, has an RNA genome that is converted into a DNA copy by the virus's own reverse transcriptase after infection of the host cell. Reverse transcriptases may be mutant forms of earlier RNA replicases. Enzymes that today replicate DNA (DNA polymerases) and transcribe it into mRNAs (RNA polymerases) may also be mutant forms of older RNA replicases or reverse transcriptases.

The RNA world hypothesis is a very attractive one because it bases the appearance of life squarely within the realm of evolution. RNA, by virtue of its ability to store information in a base sequence, propagate it by replication, and evolve through mutation, allows much more flexibility than can be seen in a world in which proteins appeared first. Indeed, even though proteins can store information through their amino acid sequences, they cannot replicate and they cannot mutate. Natural selection cannot operate on them. Not surprisingly, the most ardent proponents of a protein-first world tend to be biochemists more anxious to imagine primitive metabolic pathways than genetic information storage and evolution. Conversely, adherents to the view that the RNA world appeared first tend to be of a more genetic bent and consider that metabolic pathways are a consequence of the evolution of the RNA world. This debate is far from over, because at this point, we do not have firm evidence either way. Right now, the pendulum is swinging in the direction of the RNA world, but research on both hypotheses continues.

For how long did a putative RNA world exist? Remembering that life could not have been established before the end of the period of heavy bombardment 3.8 billion years ago, and taking into account that the first genuine prokaryotes have been dated to 3.5 billion years ago, the RNA world could not have lasted more than 300 million years—a very long time. For example, dinosaurs did not even exist 300 million years ago. At that time, Earth was populated by a large number of microscopic species and many types of sponges, mollusks, and worms. Land plants had already appeared, but not yet flowering plants. Fishes had been in existence for a while, but amphibians and reptiles were recent. However, there were no mammals and no birds, and the first hominids would not appear until about 296 million years later. Nevertheless, it is legitimate to ask whether life could have appeared from nothing in only 300 million years. Some scientists, albeit a small minority, do not think so and posit that life must have originated elsewhere. We will consider their views in the next chapter.

FIRST DNA-CONTAINING CELLS AND THEIR EVOLUTION

Microscopic fossils and their chemical footprints are not particularly rare. Fossil forms of cyanobacteria, unicellular bacteria able to perform photosynthesis, have been found in rocks dating back 3.5 billion years. This date has been contested recently, however, and cyanobacteria may be slightly more recent than formerly thought. Evidence for the existence of organic molecules that could, in all likelihood, have been synthesized only by living cells has been found even in slightly older rocks. Cyanobacteria still exist today and can form, in association with other organisms, structures called stromatolites, which are found on the coast of Western Australia, among other places. Notably, fossil stromatolites (figure 5.8) have been discovered in Ontario, Canada, and elsewhere, and they contain microscopic structures that look like modern cyanobacteria, at least as far as shape and size are concerned. Organic compounds that resemble chlorophyll degradation products have also been found in these rock formations, as well as low levels of carbon 13, a sign that photosynthetic lifeforms proliferated there a long time ago.

The element carbon exists in the form of three isotopes, carbon 12, 13, and 14. All three isotopes contain six protons in their nucleus, but carbon 12 has six neutrons, carbon 13 has seven, and carbon 14 has eight. Carbon 14 is radioactive, but its rate of decay is rather rapid, so that carbon 14 can be used only to date materials that are a few thousand years old, not billions of years old. Carbon 13 is stable and represents about 1 percent of the total carbon found as CO_2 in Earth's atmosphere, whereas carbon 12 represents 99 percent. The process of photosynthesis favors the utilization of carbon 12 over that of carbon 13 in the form of CO_2. This means that photosynthetic organisms contain low levels of

FIGURE 5.8 Fossil stromatolite from Northern Idaho. The layered structure is visible. Actual length is about 15 cm. (Courtesy of Dennis Cartwright, Washington State University.)

carbon 13, and this was exactly what was found in the rocks where cyanobacteria-like fossils were discovered. The evidence that the 3.5-billion-year-old microscopic fossils were once living cyanobacteria is thus quite good.

However, cyanobacteria are sophisticated organisms; they contain chlorophyll and their modern version can fix atmospheric nitrogen. It is not very likely, therefore, that cyanobacteria were the first DNA-containing cells. They probably evolved from a more rudimentary type of cell, called by some a progenote and by others the last common ancestor, that would have been the mother of all life-forms based on DNA. This progenote is thus placed at the very root of the global phylogenetic tree (see figure 3.5) and was itself derived from RNA-containing ribo-organisms.

What would the progenote have looked like? So far, we can only speculate. Scientists agree that it must have been an anaerobic organism, since there was still no oxygen in the atmosphere. In terms of metabolism, the progenote was probably not relying on complex nutrients (they did not exist yet, either), was not photosynthetic, and may have been a sulfur metabolizer or a methanogen, or both. Sulfur metabolizers convert elemental sulfur into hydrogen sulfide, whereas methanogens react carbon dioxide with hydrogen to form methane. Both reactions are reductions and could have been coupled with an electron transfer system to generate ATP. Several modern prokaryotic species occupy niches that are rich in sulfur, methane, and hydrogen and that are also characterized by high temperature and acidity. Thus many scientists think that the progenote originated in a hot environment, such as the volcanic ponds and springs that exist in Yellowstone National Park. What is more, these hot, volcanic gas-containing niches are populated today by several species of prokaryotes, mostly belonging to the domain Archaea.

Of course, not everybody agrees that DNA-based life had a hot origin. Recently, researchers conducted an extensive computer search of the G and C content of ribosomal RNAs from forty different organisms ranging from heat-loving bacteria to mammals. A high G+C content indicates more stability at high temperature than a high A+U content in nucleic acids. Based on a phylogenetic analysis, they found that the putative progenote was not particularly rich in G+C bases and hence may not have been a heat-loving species after all.

What does this controversy over the hot origin of DNA-based life mean? First, it means that a clashing of ideas is very much a part of the scientific process. Second, it means that we really do not yet know what the progenote truly was, and third, it may also mean that some scientists are swayed by models that seem to be in line with conventional thinking. I have alluded earlier to the fact that biology does not really have a good theoretical background against which claims and even experimental observations can be verified or rejected. A hot origin for the appearance of DNA-based life (and perhaps the RNA world as well) is consistent with the concept that Earth *was* a hot, volcanic, meteorite-bombarded place right before or during the emergence of life. This is conventional, mainstream thinking.

But we are not entirely sure that Earth was a hot place at the dawn of life. In fact, based on calculations describing stellar evolution, some scientists think that the young Sun, about 4 billion years ago, delivered only 75 percent of the thermal radiation it is delivering today. This "cold Sun" scenario would have been tempered by a strong, Venus-like but not quite so pronounced, greenhouse effect that could have made Earth hot. Some scientists dismiss this hypothesis and claim that life appeared on a glacial, "snowball" Earth. This, they say, would have much increased the stability of fragile organic molecules necessary for life to appear. What can be made of all these conflicting ideas? Not too much so far; the jury is still out.

Another problem simply has to do with vocabulary. When prokaryotes were discovered in extreme environments, such as those displaying very high salinity, temperature, or acidity, they were dubbed archaebacteria (now Archaea), in contrast to regular bacteria (now Bacteria), which inhabit friendlier niches, at least in human terms. When it was discovered that Archaea significantly differ in their molecular composition from Bacteria, the concept that they were very primitive, archaic organisms (hence their name) living under harsh conditions took root. There is no evidence that Archaea are more archaic than other prokaryotes, but the name has stuck. In fact, it is now believed that Archaea may have given rise to eukaryotes, organisms such as bread mold, spiders, goldfish, and humans. This does not make them more or less primitive than Bacteria.

One of the differences between Archaea, Bacteria, and Eukarya is the presence of D-amino acids (right-handed) in their cell wall. As we saw in chapter 4, all amino acids used for protein synthesis (even by Archaea) are of the left-handed, L type. The presence of D-amino acids in Archaea is also seen as a primitive character because the primeval broth would have contained equal amounts of L and D types of amino acids. Some books go on to say that during evolution, more "advanced" Bacteria and Eukarya lost the ability to incorporate D-amino acids in their cell envelope. However, it is equally justified to say that Archaea gained this ability. And, by the way, some members of the Bacteria *do* use D-amino acids to a limited extent to synthesize specific compounds.

These considerations simply emphasize our present ignorance of what were the progenote and its immediate descendants. One can always speculate, however. In the absence of oxygen, close descendants of the progenote must have had a metabolism of the anaerobic type. Anaerobic organisms still exist today, such as many Bacteria and Archaea, while among Eukarya, a well-known example is brewer's yeast, used to make beer. These organisms use fermentation (see chapter 3) to produce energy in the form of ATP. Descendants of the progenote also may have been able to fix atmospheric N_2 to produce ammonia, which is then converted into amino acids and nitrogenous bases. Nitrogen fixation uses the ATP formed during the fermentation process. In addition, they must have developed an electron transfer system based on porphyrin-containing compounds. Without oxygen as a reducible substrate, they may have "breathed" sulfur compounds instead to produce more ATP. Some modern prokaryotes do just that, and they use porphyrins as electron carriers. Fascinatingly, porphyrins are made in Miller-type gas-discharge experiments.

Finally, the progenote and its offspring may have developed nonoxygenic photosynthesis (also based on porphyrins), in which CO_2 is reduced into glucose, which is then fermented. Nonoxygenic photosynthesis exists today in certain microorganisms. It does not rely on the splitting of water and hence produces no oxygen. The electron donor here is not H_2O; it can be H_2S or even H_2. A recently published phylogenetic analysis of genes involved in photosynthesis strongly supports the idea that nonoxygenic photosynthesis based on bacteriochlorophyll preceded photosynthesis based on chlorophyll. Bacteriochlorophyll is found in bacteria of the green sulfur and nonsulfur types, whereas chlorophyll is found in cyanobacteria and plants. This, of course, raises the question as to why cyanobacteria and plants have the same type of chlorophyll. More about that later.

Then, the photosynthetic apparatus must have evolved in such a way that oxygenic photosynthesis became possible. This is the type of photosynthesis that cyanobacteria are capable of. The "invention" of oxygenic photosynthesis was to have dramatic consequences for the evolution of life on Earth, as we will see. The overall equation that describes oxygenic photosynthesis is

$$6\ CO_2 + 6\ H_2O + energy \rightarrow C_6H_{12}O_6 + 6\ O_2,$$

where the energy is provided by solar photons, and $C_6H_{12}O_6$ is glucose. Oxygen gas (O_2) is simply released in the environment. Cyanobacteria thus created an "oxygen crisis" and contributed to what may have been the first mass extinction afflicting many species. Prior to the appearance of oxygenic photosynthesis, life had evolved in a completely anaerobic environment. It turns out that oxygen is a very reactive gas and is in fact quite toxic to obligate anaerobes. When cyanobacteria first started evolving oxygen, it must have reacted quickly with other gases, present in solution, such as ammonia, carbon monoxide, and hydrogen sulfide, to yield nitrogen, carbon dioxide, water, and sulfur dioxide. It would also have reacted with dissolved minerals such as reduced iron. This is why not all life was wiped out. Gradually, however, as oxygen became more plentiful, it started dissolving in water and eventually escaped into the atmosphere. This spelled doom for many species of Bacteria and Archaea, and they became extinct or were relegated to oxygen-poor areas such as lake or ocean bottoms. The species that survived in the presence of oxygen could have done so only by creating oxygen-detoxifying enzymes (such as catalases and peroxidases); those that had not accomplished this became confined to anoxic niches.

Another consequence of oxygen production via photosynthesis was the progressive formation of the ozone layer. Ozone strongly absorbs the most mutagenic wavelengths of ultraviolet (UV) light, so the ozone layer would have slowed down evolution by gene mutation. However, concomitantly, the blocking of UV light enhanced the survivability of oxygen-tolerant cells living near the surface of water or on wet land. Finally, the last and perhaps most important consequence of the oxygen crisis was the appearance of aerobic respiration. As we saw in chapter 3, aerobic respiration uses oxygen gas as a metabolite to produce high levels of energy in the form of ATP. Aerobic respiration consists of a set of oxidation-reduction reactions involving specific cytochromes, all equipped with porphyrin rings. It is possible that these cytochromes evolved from those used in anaerobic respiration where metabolites other than oxygen are used as electron acceptors.

In conclusion, the evolution of cyanobacteria from the ancestral progenote and its descendants led to the appearance of two major energy-producing mechanisms that an enormous majority of eukaryotes use today: oxygenic photosynthesis and aerobic respiration. It is estimated that oxygen in the atmosphere reached its present level about 2 billion years ago. Cyanobacteria are such successful organisms that the descendants of those that oxidized Earth's atmosphere are still ubiquitous today (figure 5.9).

PATHS OF FAST EVOLUTION

The modern prokaryotic world is enormously diverse. We now know that evolution by gene mutation is but one cause of this diversity. Indeed, prokaryotes have also developed some kind of primitive sex and other methods that allow transfer of whole groups of genes from one cell to another. In all likelihood, prokaryotes living billions of years ago could do the same thing. The phenomena allowing exchange of DNA between bacterial cells are known collectively as horizontal gene transfer and comprise the mechanisms of conjugation, transduction, and transformation.

In conjugation, two prokaryotic cells become united by a tubular bridge through which the chromosome of one of the mating partners is transferred into the other cell. In this process, thousands of genes can potentially be exchanged by two mating bacterial cells. The blending of many different genes in a single cell can of course have important consequences for its evolution. The second mode of horizontal gene transfer involves DNA-containing bacteriophages as little gene carriers. When a bacteriophage infects a prokaryotic cell, its DNA is rapidly replicated many times. Concurrently, some of the bacteriophage genes are transcribed and translated to synthesize one or several coat proteins. These are destined to coat the newly formed bacteriophage DNA and so produce many new viral particles.

As the new bacteriophage DNA is being packaged by the coat proteins, DNA from the prokaryotic host can become accidentally trapped into some of

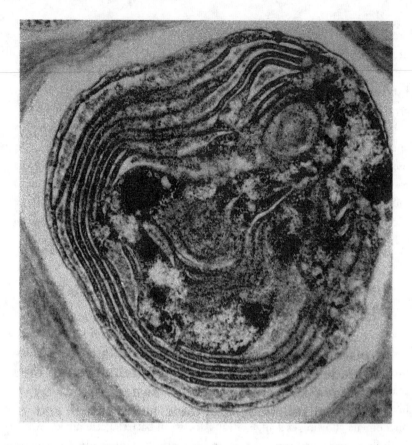

FIGURE 5.9 Electron micrograph of a modern cyanobacterium. The parallel membranes inside the cell are called thylakoids and are the sites of photosynthesis in this organism. (From Margulis, L. *Early Life*. 1984. Sudbury, Mass.: Jones and Bartlett, figure 3.1, page 57. Reprinted with permission of the publisher.)

the new bacteriophage particles. Since the amount of DNA that can be contained in the bacteriophage coat is limited, the presence of host DNA prevents trapping of bacteriophage genes. Thus bacteriophage particles containing host DNA, but no bacteriophage DNA, are formed. When the particles are released by the dying host, those bacteriophages that contain bacterial DNA (but no bacteriophage genes) can still infect a new host bacterium, but they do not kill it. The result is that the new host acquires many genes from the previously infected (and now dead) host. These additional genes blend with the existing genes of the new host and thus modify its genome.

Finally, bacterial cells die all the time in nature and, in the process, their membrane gets disrupted and their DNA is released into the environment. Some bacterial species have evolved the ability to pick up this released DNA and use it as their own to code for new functions. This phenomenon is known as transformation. The first two mechanisms of horizontal gene transfer are depicted in figure 5.10.

Clearly, horizontal gene transfer between prokaryotes can lead to very rapid evolution and diversification, since it involves the transfer of many genes at a time. Now that the genomes of dozens of prokaryotic species have been fully sequenced, it is becoming clear that such sharing of genes was quite common in the past. Prokaryotes had plenty of time to diversify; they ruled the world alone between 3.5 billion and 2 billion years ago.

FIRST EUKARYOTES

A major transition in the living world took place about 2 billion years ago: the appearance of the first eukaryotic cells. The 2.1-billion-year-old fossilized remains of a photosynthetic alga (containing chloroplasts) were discovered recently in Michigan and constitute the oldest known eukaryotic fossil. As we have seen, eukaryotic cells are very significantly different from prokaryotic cells. One of the major differences is the existence in eukaryotes of a complex cytomembrane system that includes the nuclear membrane, the cytoskeleton, and organelles such as mitochondria and chloroplasts. The cytomembrane system allows animal cells to perform endocytosis, the process by which the cell membrane creates invaginations (little pockets) in which extracellular material, including nutrients, can become trapped. These invaginations are then sealed off inside the cell and the vesicles so created fuse with preexisting cellular bodies such as lysosomes. Lysosomes contain an array of enzymes that digest whatever gets trapped in the cell membrane's invaginations.

Another important difference between prokaryotes and eukaryotes is the way in which the eukaryotic genome is organized. Bacterial chromosomes are circular, meaning that their DNA is a double helix closed on itself. By contrast, eukaryotic chromosomes are linear, and their DNA is associated with many proteins that have no counterparts in most prokaryotes. These proteins are organized together with the DNA in the form of nucleosomes (figure 5.11). Such an organization is not found in prokaryotes. Furthermore, the coding sequence

A. CONJUGATION

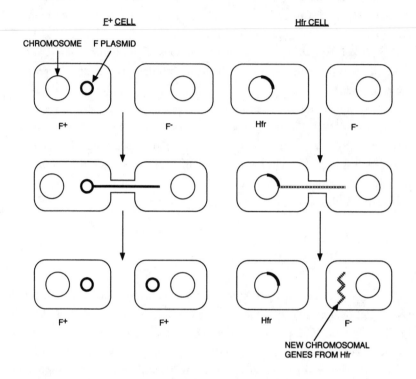

FIGURE 5.10 Schematic representation of conjugation and transduction. A: Conjugation. An F⁺ prokaryotic cell contains an F plasmid, a piece of circular DNA, in addition to its chromosome. This F plasmid harbors the genes necessary for the establishment of a bridge between an F⁺ and an F⁻ cell. Once the bridge connects the two cells, the F plasmid is transferred into the F⁻ cell. The result is the production of two F⁺ cells. Sometimes, the F plasmid becomes physically integrated within the prokaryotic chromosome, turning an F⁺ cell into an Hfr cell. An Hfr cell is able to mobilize many genes into an F⁻ recipient, thereby creating new gene combinations.

of eukaryotic genes is usually interrupted by noncoding DNA sequences called introns (figure 5.12). Introns are extremely rare in prokaryotes. Finally, there are morphological differences between the two types of cells; some of the main ones are illustrated in figure 5.13. How could all these differences have come about?

The presence of chloroplasts and mitochondria in plants cells and mitochondria in animal cells is best explained by a hypothesis developed in the late 1960s by Lynn Margulis of Boston University. Her model is known as the endosymbiont hypothesis. Margulis had been interested in the symbiotic rela-

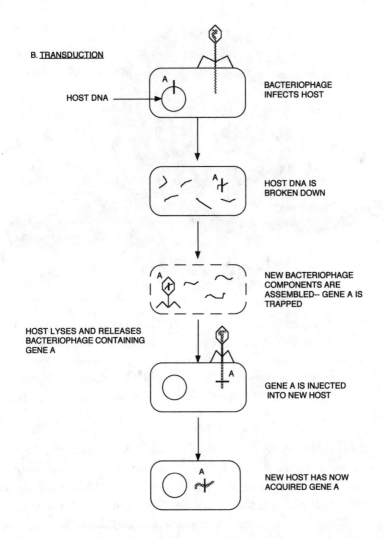

B. TRANSDUCTION

HOST DNA

A

BACTERIOPHAGE
INFECTS HOST

HOST DNA IS
BROKEN DOWN

NEW BACTERIOPHAGE
COMPONENTS ARE
ASSEMBLED-- GENE A IS
TRAPPED

HOST LYSES AND RELEASES
BACTERIOPHAGE CONTAINING
GENE A

GENE A IS INJECTED
INTO NEW HOST

NEW HOST HAS NOW
ACQUIRED GENE A

FIGURE 5.10 (Continued.) **B:** Transduction. A prokaryotic host cell is infected by a bacteriophage. In the process of bacteriophage multiplication, some pieces of host DNA (in this case, a piece carrying gene A) become packaged in newly formed bacteriophage particles. These pieces of host DNA can then be transferred to a new host cell, also creating new gene combinations. The bacteriophage and the bacterial cell are not to scale: the bacteriophage particle is about fifty times smaller than the bacterium.

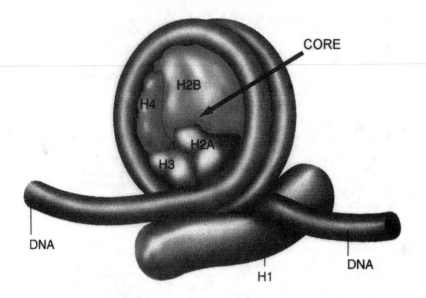

FIGURE 5.11 Organization of eukaryotic DNA in a nucleosome. DNA is represented by the coiled thin cylinder. DNA is wrapped around a core of eight histone proteins, two copies each of histones H2A, H2B, H3, and H4. Histone H1 stabilizes the DNA coil. (Adapted from Hartwell, L. H., L. Hood, M. L. Goldberg, A. E. Reynolds, L. M. Silver, and R. C. Veres. 2000. *Genetics*. New York: McGraw-Hill.)

tionships between the algae and fungi that form lichens. In lichens, these two types of organisms are intimately mixed, and they cooperate in their fight against the harsh environment in which they live. Taking this idea one step further, Margulis hypothesized that chloroplasts and mitochondria were at one time free-living prokaryotes that somehow became engulfed by larger cells and established themselves as symbionts *inside* their hosts (hence the name *endo*symbionts). At the time, this hypothesis was greeted with a good deal of skepticism. Now that the properties of chloroplasts and mitochondria are much better understood, Margulis's hypothesis has become a classical model presented in elementary biology textbooks.

Indeed, we now know that both mitochondria and chloroplasts contain their own DNA. The general architecture of their genomes, their protein-synthesizing apparatus, and their sensitivity to certain antibiotics make it clear that they are of prokaryotic origin. Mitochondria and chloroplasts need the cooperation of nuclear genes to function properly because, over time, many genes they orig-

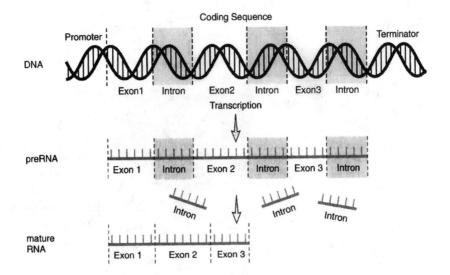

FIGURE 5.12 Introns in eukaryotic genes. A eukaryotic DNA gene contains coding (exons) and noncoding (introns) base sequences. Both exons and introns are copied into an mRNA molecule (preRNA), which is then processed for intron removal. After intron elimination, exons are stitched together to form an uninterrupted coding sequence consisting of exons only (mature RNA). (Adapted from Alcamo, I. E. 1996. *DNA Technology: The Awesome Skill.* Dubuque, Iowa: Wm. C. Brown.)

inally housed have been transferred to the nucleus of plant and animal cells. Thus in the endosymbiont model, animal cells presumably originated from the capture of aerobic bacteria—possibly from a class known as purple bacteria and capable of making ATP via respiration—by larger anaerobic bacteria able only to ferment. Plant cells, in turn, came from large anaerobic prokaryotes that had engulfed the progenitors of mitochondria, as well as from cyanobacteria and their photosynthetic machinery that later on became chloroplasts. In both cases, the evolutionary advantage of endosymbiosis was a vastly greater ability to produce ATP. Furthermore since all eukaryotes that possess chloroplasts also possess mitochondria, many think that acquisition of mitochondria came first.

Curiously, it has been recently demonstrated that the eukaryotic human pathogen *Toxoplasma*, which is not a plant, contains an organelle that looks very much like a remnant from a unicellular photosynthetic organism! This organelle contains DNA that is related to that of a green alga (also a eukaryote, not a prokaryote like a cyanobacterium) but that has no chlorophyll and has lost

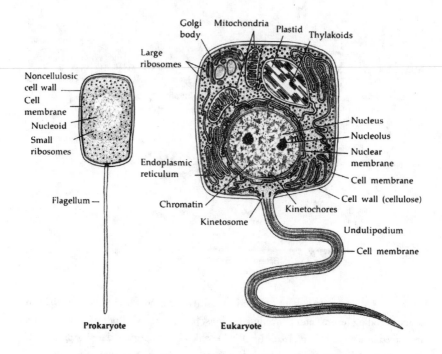

FIGURE 5.13 A summary of the main morphological differences between prokaryotic and eukaryotic cells. The eukaryotic cell is a composite containing features not necessarily present in all eukaryotic cells. Differences and degree of complexity are obvious. (From Margulis, L. *Early Life*. 1984. Sudbury, Mass.: Jones and Bartlett, figure 1.1, page 3. Reprinted with permission of the publisher.)

the genes responsible for photosynthesis. The explanation for this is that this organelle once was a free-living eukaryotic cell engulfed by another eukaryotic cell. Thus this capture must have occurred more recently than the ones that gave rise to the first eukaryotes. Further study of *Toxoplasma* may shed light on the phenomenon of endosymbiosis in general.

Although the endosymbiont model is now well accepted by the scientific community, the problem of identifying the hosts that captured purple bacteria and cyanobacteria to produce the first eukaryotes is not solved. What could these hosts have been? Not all eukaryotes contain organelles. For example, the unicellular human parasite *Giardia lamblia* has neither mitochondria nor chloroplasts. Could it be that the ancestors of *Giardia* were primeval eukaryotes that never captured any endosymbiotic guests, in contrast to most other eukaryotes? This seems unlikely, because there is good evidence that the nuclear genome of *Giardia* contains genes of prokaryotic origin. This evidence suggests that the ancestors of *Giardia* once contained organelles of prokaryotic descent that transferred some of their genes to their nucleus and then disappeared. *Giardia* is thus not a good example of a "primeval" eukaryote. Some other answers may be found when the genomes of more unicellular eukaryotes, with and without mitochondria or chloroplasts, are sequenced.

In the meantime, scientists have formulated a more intricate hypothesis to explain the origin of the first eukaryotic cells. This new model is called the hydrogen hypothesis because it assumes that eukaryotes originated from the fusion of an anaerobic, hydrogen-dependent archaeal cell with an aerobic bacterial cell that was able to respire oxygen (to produce CO_2) and that also produced hydrogen as a waste product. The evolutionary advantage of this fusion would have been that the two partners, now together, could have made use of waste products (CO_2 and H_2) and could have benefited from plentiful ATP produced by respiration.

This hypothesis rests on the following observations. First, some modern eukaryotes, devoid of mitochondria, contain an organelle, called the hydrogenosome, that produces hydrogen gas and ATP. The hydrogenosome contains no DNA (which may have been lost or transferred to the nucleus), but some of its components resemble that of mitochondria. This suggests a bacterial origin for these hydrogenosomes. Furthermore, many contemporary Archaea strictly depend on H_2 and CO_2 to produce ATP. Thus an ancestral archaeal cell could have had a similar metabolism. Finally, phylogenetic analysis shows that eukaryotes are more closely related to Archaea than they are to Bacteria (more about that later). Therefore saying that eukaryotes descend from *both* Bacteria and Archaea makes a lot of sense (figure 5.14).

This hypothesis is falsifiable, because if correct, it means that eukaryotes harboring mitochondria have kept the respiration function (oxygen metabolism) of the original bacterial portion of the fusion but have lost the ability to generate hydrogen. On the other hand, in eukaryotes devoid of mitochondria but able to produce hydrogen, it is the respiration function that has been lost and the ability to generate hydrogen that has been kept. Finally, for those eukaryotes that contain neither mitochondria nor hydrogenosome, it can be hypothesized that many of the bacterial genes were transferred to the archaeal cell genome (presumably to provide some needed functions) before the bacterial portion of the partnership disappeared. Again, sequencing of a variety of eukaryotic genomes will either support or refute this hypothesis. In all cases, traces of ancestral genes, responsible for respiration or hydrogen production or both, should be found in eukaryotes that can perform only one of these two functions

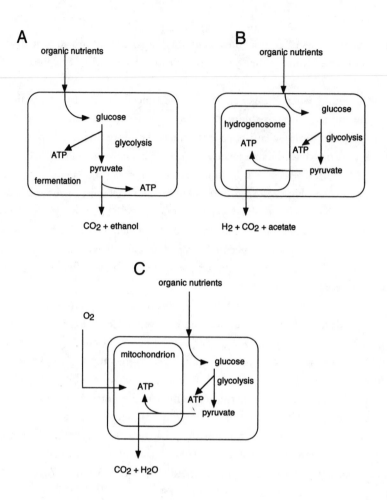

FIGURE 5.14 The hydrogen hypothesis. Three modern outcomes of the hypothetical fusion between an archaeal cell and a bacterial cell are shown. **A:** An amitochondriate anaerobic eukaryote, such as *Giardia*, produces ATP by glycolysis and fermentation. In some cases, lactate is produced instead of ethanol. Both the ability to produce hydrogen and the ability to perform respiration have been lost. **B:** In these anaerobic cells, the hydrogenosome represents the remnant of the aerobic bacterial cell that originally produced hydrogen as a waste product. The respiration function has been lost. **C:** This mitochondriate aerobic cell has lost the ability to produce hydrogen but has kept the respiration function present in the mitochondrion. All three types of cells have kept the anaerobic glycolytic pathway inherited from the archaeal fusion partner. (Adapted from Martin, W., and M. Muller. 1998. The hydrogen hypothesis for the first eukaryote. *Nature* 392:37-41.)

or none at all. We do not have answers yet because, for now, genome sequencing efforts are restricted to the human genome and that of model organisms such as the mouse, the rat, and the fruit fly.

The hydrogen hypothesis, however, does not explain how the bacterial-archaeal partnership developed a cytoskeleton and an intricate membrane system. Indeed, neither Bacteria nor Archaea have a cytoskeleton or a complicated membrane system. Not enough is known yet about the genetics of the cytoskeleton and eukaryotic membranes to make educated guesses about their origin. It is interesting to note that some eukaryotes have a very simple membrane system. Also, eukaryotes possess in their membranes some protein and lipid components that seem to be of archaeal origin. As for the cytoskeleton, its origin is presently anyone's guess. Clearly, much more research is needed in these areas.[4]

Other features of eukaryotic cells that the hydrogen hypothesis does not explain are the existence of linear chromosomes (as opposed to circular bacterial and archaeal chromosomes) and the ability to "cap" molecules of messenger RNA. Indeed, unlike prokaryotic mRNAs, eukaryotic mRNAs always contain a modified guanine as the very first base. In addition, eukaryotic mRNAs are terminated with a "tail" of many adenines strung together. The function of the modified G is to firmly dock eukaryotic mRNAs to ribosomes just before translation begins. Prokaryotes use a completely different mechanism to attach their mRNAs to ribosomes. The function of the tail is to provide eukaryotic mRNAs with protection from degradation. Since eukaryotic cells do not divide faster than once in at least 24 hours, their mRNAs must display greater stability than prokaryotic mRNAs, whose hosts divide much faster.

What is the origin of these three features? It turns out that many modern viruses harbor a linear DNA chromosome and also cap and tail their mRNAs. Hence, an even newer hypothesis, published in 2001, now proposes that all eukaryotes are ultimately derived from an ancient virus that fused with an archaeal mycoplasma. A mycoplasma is a prokaryotic cell that is devoid of a cell wall; it is able to undergo membrane fusion with other organisms simply because there is no cell wall to interfere with this fusion mechanism. Modern archaeal mycoplasmas do exist and may descend from very old ancestral forms.

What is more, many viruses possess an outer lipid membrane that covers their protein capsule. These viruses penetrate modern cells by fusing with the cellular membrane. Therefore the new viral hypothesis contends that such a virus invaded an archaeal mycoplasma host, where it became established and became the proto-eukaryotic linear chromosome by recruiting genes from the archaeal chromosome. Many archaeal genes were subsequently lost by this chimeric organism. But then, we also know that modern eukaryotes contain the descendants of many bacterial genes. Where are these coming from? Well, since the ancestral mycoplasma—now containing a viral chromosome—did not have a cell wall, it may also have been able to perform endocytosis and may thus have engulfed bacterial prey whose genes were then marshaled to perform new, useful functions. If correct, the virus hypothesis surmises that viruses are quite old, as old as prokaryotes. We do not actually know this to be the case. At any rate, the virus hypothesis would be supported by the discovery of complex viruses capable of infecting archaeal mycoplasmas. Such viruses have not yet been discovered. Finally, the virus hypothesis and the hydrogen hypothesis should not be seen as contradictory. Rather, it can be argued that the mycoplasma host engulfed carbon dioxide- and hydrogen-producing bacteria (as in the hydrogen hypothesis) and in addition, it also acquired permanent DNA viruses that ended up making the linear eukaryotic chromosome (figure 5.15).

The last issue I wish to discuss in this chapter is the origin of the complex genome organization in eukaryotes. As we saw in figure 5.11, eukaryotic DNA is wrapped around a core of histone proteins. There are five such different histone proteins in all eukaryotes. Interestingly, five histone genes have been identified in the genome of the archaeal *Methanococcus jannaschii*, a prokaryotic species that dwells near hydrothermal vents. But surprisingly, the genome of this Archaea is not organized in nucleosomes; the function of the histones in this organism is unknown. What is more, the fundamental genes coding for DNA replication and transcription in Archaea are very similar to the corresponding genes found in Eukarya. This adds much weight to the hypothesis that eukaryotes are more closely related to archaeal cells than they are to bacterial cells. Again, the term *archaeal* (meaning archaic) should not be taken literally in this context, because Archaea look in some ways less archaic (at least in a human idiosyncratic framework) than Bacteria. The great obscuring feature of all this research on the origins of eukaryotes is that archaeal genes are found in bacterial cells and vice versa. This is in all likelihood the result of horizontal gene transfer that probably occurred eons ago (and very possibly still occurs today) among prokaryotic species. Potentially, the sequencing of many more prokaryotic and eukaryotic genomes will provide clues to the difficult question of the origin of eukaryotes.

Finally, and again concerning eukaryotic genome organization, there is the question of the origin of introns. Introns, DNA sequences that interrupt the coding sequences of eukaryotic genes, are extremely rare in prokaryotes, but they are not absent. In addition, chloroplast genes, thought to be of ancient

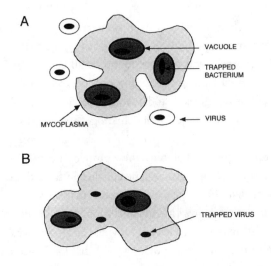

FIGURE 5.15 The virus hypothesis. **A:** An archaeal mycoplasma with bacterial endosymbionts trapped in vacuoles. Virus particles surround this cell. **B:** A mycoplasma, with endosymbionts, stably infected by a complex DNA-containing virus destined to become the eukaryotic chromosome. After fusion with the mycoplasma membrane, the virus particles have lost their lipid membrane. (Adapted from Bell, P. J. L. 2001. Viral eukaryogenesis: Was the ancestor of the nucleus a complex DNA virus? *Journal of Molecular Evolution* 53:251-256.)

bacterial ancestry, do contain introns. The question then is, have introns been there from the beginning in prokaryotes (or even before that, in the RNA world) and were they subsequently lost by the enormous majority of prokaryotes? On the other hand, are introns a new, eukaryotic "invention" that found its way to the genomes of rare, contemporary prokaryotes?

Two hypotheses have been formulated. One holds that introns are ancestral and assumes that they may have originated in the RNA world. In this model, introns are seen as nucleic acid sequences, devoid of function and present by mistake in original RNA genes. These noncoding sequences then allowed exon (the coding part of genes) shuffling and creation of new genes. Indeed, the presence of noncoding sequences in ancestral genes would have allowed "cutting and pasting" of exons to create a great diversity of proteins during evolution. These introns would then have been largely lost by most prokaryotes and a few unicellular eukaryotes (*Giardia*, for example, does not seem to have introns in its genes). In the second hypothesis, introns are simply a property of the eukaryotic world. Which is right?

If the exon shuffling model is correct, some argue, it should be possible to recognize some common patterns in ancient genes (genes that may have developed in the RNA world and that are found in all cells today). These genes would have originated from the assembly of old exons, and these blocks of DNA should still be recognizable today in these ancient genes. No pattern was found in the genes coding for alcohol dehydrogenase, globins (proteins carrying the ancient porphyrin rings), pyruvate kinase, and triose phosphate isomerase. With the exception of globins, these proteins are involved in the ancestral glycolysis/fermentation pathway. The conclusion from these results is that introns are not ancient. Of course, one can retort that the study of four genes is much too limited to reach any kind of conclusion. Additionally, the opponents of the "old introns" hypothesis have not provided an alternative explanation for their origin. In conclusion, we do not know with any kind of certainty what the origin of introns is.

CONCLUSIONS

As you are now aware, the science of the origins of life is very challenging. Exciting hypotheses have been formulated and crisp mathematical models have been developed, yet enormous uncertainties remain at practically every step of the way. Are we getting closer to an answer? This is impossible to predict, as is usual in science. More research, in particular the full sequencing of hundreds of genomes, may give us more clues as to their ascent. This will take some time, however, because sequencing efforts are presently restricted to organisms of medical or agronomic importance. Or we may find answers not on Earth but elsewhere in the solar system, if extant life or traces of extinct life exist on other planets and satellites.

The evolution of life on Earth clearly proceeded well beyond the appearance of eukaryotic cells. We are here to prove it! However, it has not been my intention to describe in this book the full path that life took in the 2 billion years that followed the creation of the first complex cells. First, this would be a daunting task, and second, what happened after 2 billion years ago seems a little more mundane than the mystery of the origins. (I can already hear the cries of protest!) Nevertheless, in many ways, what happened after the first eukaryotic cells appeared is relatively easy to explain. In a nutshell, multicellularity was achieved, then cell differentiation into different organs burst into existence, fol-

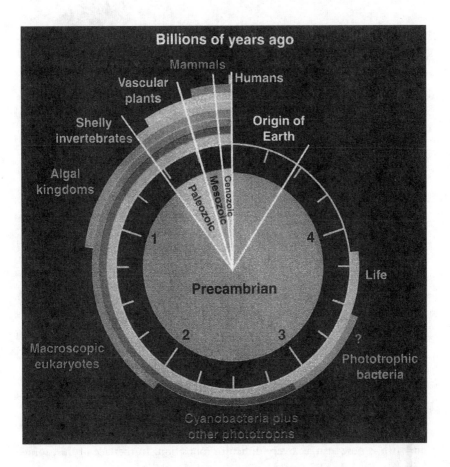

FIGURE 5.16 A grand summary of the origin and evolution of life on Earth. (From Des Marais, D. J. 2000. When did photosynthesis emerge on Earth? *Science* 289:1703-1705, courtesy of the American Association for the Advancement of Science.)

lowed by the evolution of sex, animals, plants, and humans. This took only about 1 billion years, a drop in the universe's time bucket and only about one third of the time it took Earth to become host to unicellular eukaryotic life (figure 5.16). Furthermore, fossils of multicellular organisms are plentiful and easy to date, and the modern animals, plants, and prokaryotes that people study are right at hand. What cushy jobs paleontologists and molecular biologists have! Certainly, they will disagree. Any other reaction would not, after all, be expected from scientists.

CHAPTER 6

Has Life Originated Elsewhere and Will It End?

If the genetic code is universal, it is probably because every organism that has succeeded in living up till now is descended from one single ancestor. But, it is impossible to measure the probability of an event that occurred only once.

—FRANÇOIS JACOB, *The Logic of Life* (1973)

Those who claim that life is a highly improbable event, possibly unique, have not looked closely enough at the chemical realities underlying the origin of life. Life is either a reproducible, almost common-place manifestation of matter, given certain conditions, or a miracle.

—CHRISTIAN DE DUVE, *Vital Dust* (1995)

The wave of optimism that followed the Miller experiment in 1953 has been replaced today by more subdued attitudes. Solving the mysteries of the emergence of life on Earth is now seen as a very difficult proposition. In the face of this, some have proposed that life may have originated elsewhere in the solar system, or even somewhere else in our galaxy. Proponents of this view include the famous biophysicist Francis Crick of DNA fame (his statement on that question seems to have been tongue-in-cheek) and the equally famous cosmologist Fred Hoyle (not so tongue-in-cheek). But others, nonscientists I must add, have turned this hypothesis of extraterrestrial origin of life into an absolute certainty, as the following bizarre story shows.

On 13 December 1973, the French journalist and one-time race car driver Rael was personally contacted by an Elohim, an advanced being from another planet, who asked him to establish an embassy for him and his fellow aliens on Earth. According to the Raelian home page, "The extra-terrestrial was about four feet in height, had long dark hair, almond shaped eyes, olive skin and exuded harmony and humour." He told Rael, "We were the ones who made all life on Earth, *you mistook us for gods*, we were at the origin of your main religions. Now that you are mature enough to understand this, we would like to enter *official contact* through an embassy" (emphasis theirs).

The Elohim message further announces that life on Earth is not the result of random evolution, nor the work of a supernatural God. It is a deliberate creation, using DNA, by a scientifically advanced people who made human beings in their image (except most of us are taller than the Elohim Rael met). Raelians call themselves "scientific creationists" and claim that the Elohim maintained contacts with humans via prophets such as Buddha, Moses, Jesus, and Mohamed, all selected and educated by them. In the same vein, the Virgin Mary was abducted and artificially inseminated by the Elohim, aboard their spaceship. Now that we are ready to understand their message, the Elohim want to show us the way to enlightenment, a better society, love and pleasure, human rights, and respect for one another's differences.

The goals announced by the Elohim seem reasonable enough and not unlike those to which most of us would aspire. Moreover, the origin-of-life model presented by the Raelians (there are hundreds of thousands of them) is self-consistent. It explains why all living creatures are based on DNA, to say nothing of solving the difficult question of the Virgin Birth of Christ! Furthermore, we now understand why there were prophets. But one nagging question remains: why did the Elohim talk to Rael and not to the United Nations or to the President of the United States (as in the movie *The Day the Earth Stood Still*)? Why did they not commandeer television stations to let us know all at once that they had descended upon us?

Regardless of what Rael does (fund raising is one activity) and what he thinks, one must also wonder who or what made the Elohim. Since they are not gods, they could not have made themselves. Since, they claim, there are no gods, gods could not have made them, either. Are the Elohim thus the result of natural selection on an extraterrestrial planet? They are not telling us. Therefore Rael's explanation for the origin of human life is a circular argument. Be that as it may, the Elohim (and others like them) raise the question of terrestrial life having originated elsewhere. This question is far from new and vastly antedates Rael. In the scientific community, it continues to be raised, and we will see some examples in this chapter.

PANSPERMIA

In the nineteenth century, the well-known Swedish chemist Svante Arrhenius hypothesized that a lifeless Earth was seeded with life-forms from outer space. Unlike Rael, however, he did not think that all creatures originated from clever extraterrestrial genetic engineers. Arrhenius's life-forms were simply bacterial spores that drifted through space, landed by chance on Earth, germinated, and started the process of evolution. Arrhenius called this the concept of panspermia, suggesting that maybe life originated somewhere in our galaxy and then became distributed through it. This was an interesting idea, and it solved the problem of the origin of life on Earth.

As expected, that idea was criticized from several angles. Granted, spores are hardened, dehydrated, dormant cells formed by some prokaryotic species. They are more resistant to heat, cold, and radiation than dividing cells. Spores are known to be the hardiest forms of life on Earth and some have been revived after spending 100,000 years in their dormant state. A somewhat controversial report even claims that 25- to 40-million-year-old spores have been revived in the laboratory. Spores contain a unique enzyme that can repair ultraviolet (UV)-damaged DNA very efficiently, they are resistant to temperatures up to 150°C, and they can withstand pressures as high as 6000 atmospheres and as low as 10^{-11} atmosphere. They are also quite resistant to gamma radiation.

However, are they hard enough to have resisted conditions prevalent in outer space? Assuming bacterial spores were ejected into space by gigantic volcanic eruptions or asteroidal impacts on a solar or on an extrasolar planet, which is possible, it would have taken them perhaps millions of years, depending on the distance between another solar planet and Earth or between their star and the Sun, to reach our planet. Meanwhile, the spores, even if hidden in the midst of dust grains or chunks of rock, would have been exposed to hard cosmic and ultraviolet rays that pervaded interstellar space. How long could they have resisted?

Experiments conducted in the laboratory on Earth and aboard spacecraft, mostly with the spores of the soil bacterium *Bacillus subtilis*, show that panspermia within the solar system and even interstellar panspermia are valid concepts. As we have seen, laboratory experiments have demonstrated the high tolerance of bacterial spores to injuries inflicted by temperature, radiation, and extremes of pressure. Basically, panspermia is possible only if (a) spores can be ejected from another planet and resist the acceleration and heat produced in the escape process, (b) spores can tolerate space travel for perhaps long periods of time, and (c) spores can survive the reentry process into Earth's atmosphere.

The escape process must have involved accelerations in excess of the host planet's gravity. Laboratory experiments have shown that spores can tolerate up to $460,000 \times g$ of acceleration, which is plenty to reach escape velocity from terrestrial types of planets. Then, spores must have been able to survive the hostile environment of space during their travel, during which they would have been exposed to stellar protons, electrons, alpha particles, cosmic rays composed of heavy ions, UV radiation, and X rays. They would also have experienced the effects of a high vacuum and extremes of temperature.

Several satellite-based experiments have shed a considerable amount of light on the hardiness of spores in outer space. These experiments were conducted by the National Aeronautics and Space Administration (NASA) on Spacelab, the Long Duration Exposure Facility, and Apollo missions 16 and 17; by the European Space Agency aboard the European Retrievable Carrier; and by the Russian Foton spacecraft. The experiments demonstrated that 30 to 80 percent of *B. subtilis* spores survived 6 years of exposure to vacuum and radiation in outer space near Earth if they were embedded in crystals of glucose or salt, imitating meteoritic rock. Only a thin layer of this material was needed to completely shield the spores from UV radiation. Experiments conducted outside the Van Allen belts—which deflect charged particles, including most cosmic rays, from Earth—have shown that protection from cosmic rays is more challenging. Even so, it has been calculated that some spores would survive up to 25 million years in deep space if shielded by 2 to 3 meters of meteoritic material. The notion of spores embedded in meteorites naturally raises the question of whether bacteria can live in rock. The answer to that question is yes, as living bacteria have been found in boreholes as deep as 2.5 km inside Earth's crust.

Next, there is the problem of the entry of these putative spore-bearing meteorites into Earth's atmosphere. Conventional wisdom dictates that they should burn up. After all, shooting stars are small meteorites that meet a fiery death upon entry. However, it turns out that a hot (incinerating) entry is not necessarily the norm for all small meteorites, as long as their mass exceeds about 1 kg. Examination of a Martian meteorite[1] has revealed that its interior did not exceed a temperature of 40°C, from the time of its ejection from Mars until its arrival on Earth. This temperature is totally compatible with life. The same could hold true for all sorts of meteorites, even of extrasolar origin, if they exist.

In conclusion, life could have originated elsewhere in the solar system and could have been imported to Earth. In fact, the 6 years spent in orbit by the Long Duration Exposure Facility correspond to a trip from Saturn to Earth. The interstellar panspermia concept seems more far-fetched but not totally impossible. It has been calculated that extrasolar meteorites could make the trip to the solar system in 10^5 to 10^6 years, a long time to spend in deep space, but a time still compatible with spore survival, provided they are well shielded by rocky material.

Thus the idea of panspermia has not been abandoned, even though the experiments just described can never prove that it happened. As we have seen, some distinguished scientists think that it is still a valid hypothesis, in particular in light of all the great difficulties associated with the production of organic molecules on prebiotic Earth and the relatively short duration of an RNA world. Panspermia does not, however, answer the basic question, if life did not appear from nothing here on Earth, how did it happen elsewhere? But perhaps we should determine first *if* it could have happened elsewhere.

The fact that the interior of a Martian meteorite remained relatively cool upon entering Earth's atmosphere has given some the idea that microorganisms could have traveled from Mars to Earth and perhaps seeded it. This idea is not quite as general as Arrhenius's, but it is more reasonable given the closeness of the two planets. Some scientists claim that since the transit from Mars to Earth could in rare cases take less than a year, cosmic ray exposure would not be a problem. But then, the seeding of Earth by Mars depends on the presence of past or present life on that planet. Is there, or was there, life on Mars and perhaps elsewhere in our star system?

LIFE ELSEWHERE IN THE SOLAR SYSTEM?

Until recently, this question would have been answered in the negative with a reasonable degree of certainty. Mercury and Venus are too hot; Mars and the gas giants are too cold. This attitude changed considerably after Mars and the satellites of Jupiter came under close study. Mars in particular has always exerted great public fascination because of its changing appearance with Martian seasons and the erroneous discovery of "canals" on its surface.[2] Science fiction books and movies have often depicted Mars as a dying planet where the last individuals of a technologically advanced society make nefarious plans to invade Earth. Unfortunately (or fortunately), our spacecraft have not detected any signs of an advanced society or even any clear-cut signs of life on Mars. However, some scientists think that life may once have existed there and may still exist today.

The Viking landing crafts that explored Mars in 1976 were equipped with life-detection instruments. Most of the results turned up negative, and one positive result was attributed to unusual Martian soil chemistry. At best, these results were inconclusive. Then, in 1996, a group of NASA scientists made the astonishing declaration that they had detected fossil life in a Martian meteorite. For this they did a microscopic and chemical study of Martian meteorite ALH84001, which had landed in Antarctica in 1984. In the meantime, Mars orbiters had given plausible evidence that at one time, flowing water and even shallow oceans were present on a younger, warmer Red Planet. The stage was set for the discovery of life on Mars. And this life, albeit in fossil form, was discovered! Or was it? Here is what the NASA scientists reported.

ALH84001 is a 4.5-billion-year-old piece of Mars. It has fractures where small magnesium, iron, calcium, and manganese carbonate globules are found. Interestingly, the ratio between carbon 13 and carbon 12 in these globules is in favor of carbon 12, a sign of biological activity, as we saw in an earlier chapter. In addition, the globules contain microscopic crystals of an iron oxide, called magnetite, that look exactly like the ones that some terrestrial microbes manufacture. Furthermore, the meteorite contains organic molecules of the polycyclic aromatic hydrocarbon (PAH) type. And finally, microscopic analysis revealed that ALH84001 contains small elongated bodies that look like fossil bacteria found on Earth. Taken together, these results provide strong support for past microbial life on Mars. But again and again, there are problems. Most of the fossil forms are much smaller than bacteria found on Earth today and may be too small to house all the ingredients necessary for life. Next, PAHs are abundant in space (see chapter 4) and could have been included in the rocks as the Martian surface formed. Finally, the carbon-13-to-carbon-12 ratio is not skewed enough to be totally convincing. Thus many think that the features found in the meteorite cannot be attributed to life. In addition, in 2002 researchers reported that iron carbonate in rocks could be turned into microscopic magnetite crystals through the action of heat, thereby precluding a necessary biological origin for these crystals. Thus the question of past life on Mars remains open.

Whether Mars harbored life in the past or harbors it today can be decided only by going there and digging into the Martian soil. Many aspects of the Martian surface are best explained by the past existence of liquid water (and perhaps life), and some even argue that liquid water may exist today below the permafrost. Only specific robotic or manned missions will finally tell us if life appeared on our neighbor planet. Needless to say, even if Mars is lifeless now, signs of past biological activity would be thrilling.

No scientist would have the intestinal fortitude to publicly claim a firm conviction that life exists on the giant gas planets. However, in his best-selling book

Cosmos, Carl Sagan, supposedly in one of his jocular modes, fantasized about creatures inhabiting the atmosphere of Jupiter. He imagined two types of life-forms: prey and predators, both airborne, called floaters and hunters, respectively. The floaters are balloon-like creatures propelling themselves in the manner of an octopus, except they do it with gas. They feed on the rich mixture of organic molecules produced by lightning passing through hydrogen, methane, ammonia, and water, which are known to be present in the atmosphere of Jupiter. The hunters possess streamlined bodies and vast pointed wings that look like large shark fins. They feed on herds of innocent floaters. Sagan, unfortunately, did not speculate on the origin and evolutionary history of these creatures, but by telling the story of hunters and floaters, he was trying to point out that strange environments could see the emergence of life. Interestingly, a strange environment, although much less bizarre than the atmosphere of Jupiter, has been found in the solar system; it goes by the name of Europa.

Europa is one of the four large satellites of Jupiter discovered in 1610 by Galileo. The other three Galilean satellites are called Ganymede, Io, and Callisto. Europa is about the size of the Moon and has a surface temperature of $-163°C$ at the equator. Until recently, all four satellites were of secondary interest compared with their parent planet, the largest one in the solar system. This view changed considerably when the Voyager space probes took a closer look at the satellites in the 1970s. Io, the innermost satellite, showed large volcanic plumes containing vast amounts of elemental sulfur. Europa, the second satellite, showed a strange, mostly white surface with very long, red-brown cracks. Earlier telescopic observations had shown that the white material was solid water (i.e., ice). However, the striations were a new discovery, and they suggested that perhaps floating plates of ice had moved on top of liquid water, collided with one another, and forced a red-brown solution of dissolved material to the surface, where it froze (figure 6.1).

Twenty years later, the Galileo spacecraft returned to the Jovian system and studied Europa much more closely than the Voyagers had. High-resolution images of the European surface confirmed the suspicion that moving ice plates had created the rifts, now thought to be filled with a frozen solution of magnesium sulfate (or Epsom salt) and sulfur or iron compounds. Furthermore, measurements of the magnetic field of Europa can be best reconciled with the presence of a global, salty ocean beneath the frozen surface. It is estimated that the mean depth of this ocean is about 100 km and that the ice crust is a few kilometers thick.

FIGURE 6.1 Image of Europa taken by Voyager 2 at a distance of 246,000 km. Long cracks in the surface are readily visible. (Courtesy of the National Aeronautics and Space Administration.)

How does one explain the existence of liquid water at such a great distance from the Sun? The answer resides in the tidal flexing that the Galilean satellites are experiencing as they orbit Jupiter. The combined gravitational pulls of Jupiter and the ones the satellites are exerting on one another create a push-pull effect that distorts them, much like the flexing of a tennis ball. This flexing creates heat that, on Io, results in the melting of rock and huge volcanic eruptions. On Europa, where the effect is less pronounced, the heat generated can keep water liquid underneath the crust. A similar global ocean is also believed to exist on Ganymede, but there, the ice crust is thought to be much thicker than on Europa.

Liquid water is a prerequisite for the existence of life. Is there life on Europa and perhaps on Ganymede? We do not know. Again, we will have to go there to find out. We hope that in the next few years a robotic probe equipped with an ice penetrator will be launched toward Europa to look for the presence of life. If life or its building blocks are discovered on Europa, we will have taken a giant step toward understanding the beginning of life on Earth.

As we saw earlier, space chemistry is varied and rich. Another object in the solar system that has attracted the attention of scientists interested in the origin of life is Titan, the very large satellite of Saturn. Titan is as big as the planet Mercury. At $-179°C$, its surface temperature is even lower than that of Europa. Titan is large enough to have an orange-red atmosphere, so thick that it hides the surface. This atmosphere is made mainly of nitrogen, methane (CH_4), and argon. It is thought that the surface of Titan is made of water ice covered by liquid methane, ethane (C_2H_6), and nitrogen. Nobody has ever considered that liquid water, and hence life, could exist on Titan today. However, its atmosphere could one day be used as a natural laboratory to look for the building blocks of life. Gas sparking experiments with a simulated Titan atmosphere have yielded various tarlike compounds composed of chains of hydrocarbons and hydrogen cyanide. When mixed with water, these compounds produce amino acids and nitrogenous bases.

There most probably is no liquid water on Titan today, but there may have been some much earlier, during the period of heavy bombardment of the solar system. In that case, the first steps toward life may have been taken on Titan as well. Here also, missions to this satellite should reveal the presence or absence of extraterrestrial biogenic compounds. The Cassini probe, now en route to Saturn and Titan, may bring some surprises for us.

Finally, our search for extraterrestrial life should not be confined to the solar system. As we saw in chapter 2, a significant number of stars in our corner of the galaxy have planetary companions. Because these planets are not yet directly observable, we do not know whether any of them harbor life, so we do not know their chemical composition or surface temperature, or whether some of them possess an atmosphere. This situation will change when we launch next-generation space telescopes that can visualize these planets and do spectroscopic measurements on them. For example, detection of ozone in the atmosphere of an extrasolar planet would be the signature of oxygen gas and, in all likelihood, a sign of the existence of life-forms responsible for the production of this oxygen. Unfortunately, the enormous distances separating us from even nearby stars preclude (probably for a long time to come) even uncrewed missions to these star systems.

But why should we restrict our search to *nearby* stars? Why not explore the galaxy and even the whole universe for the presence of life? Difficulties increase enormously as distance increases. Imaging planets orbiting stars located thousands of light-years away is not yet technically feasible. Nevertheless, scientists have recently proposed the concept of a galactic habitable zone (GHZ). This concept extends the idea of a circumstellar habitable zone (CHZ) to a whole galaxy. The CHZ is a region of space around a star where life may exist on orbiting planets, and its size is determined by temperatures compatible with the persistence of liquid water for a few billion years, as well as other factors. For example, the Sun's CHZ starts after the orbit of Venus and stops approximately at the orbit of Mars.

The GHZ defines the places in our galaxy that are most hospitable to life. It forms a ring around the galactic center, neither too close nor too far from it. Stars that are far from the galactic center show a low abundance of elements heavier than hydrogen and helium. These elements are, of course, critical for the formation of planets and life as we know it. Hence, without planets and heavy elements, life is extremely unlikely to have developed in the outer reaches of the galaxy.

Closer to the galactic center, star density increases and so does the abundance of heavy elements. This means that Earth-like planets could form, but high levels of radiation caused by the crowding of stars would make these planets inhospitable to life. Furthermore, it is now believed that many star systems are surrounded by cometary halos. (This halo is called the Oort cloud in the case of the solar system.) Given the proximity of stars in the inner regions of the galaxy, these halos would easily be gravitationally disturbed, resulting in the frequent bombardment of these extraterrestrial planets. These conditions are also incompatible with life. In summary, the GHZ concept strongly restricts the appearance and persistence of life in the galaxy.

Clearly, our current technological achievements do not yet enable us to search directly for the existence of inanimate life-forms outside of the solar system. But what about *intelligent* life in the universe? Receiving signals, such as radio signals, from an alien civilization would, of course, prove once and for all that life exists or has existed elsewhere in the universe. The search for extraterrestrial intelligence (SETI) had its start in 1960. That year, American astronomer Frank Drake pointed the Green Bank (West Virginia) radiotelescope at two nearby stars, Epsilon Eridani and Tau Ceti, located about 11 light-years from Earth. Not surprisingly, no artificial radio signals were detected. In terms of financial support, SETI has had ups and downs over the years, but a small group of dedicated radioastronomers continues to survey the sky for putative extraterrestrial civilizations. Funding is not the only problem facing SETI. Most astronomers are reluctant to spend much effort on a project that may yield no results at all for a lifetime. What if ET (as the extraterrestrial being in the movie of that name was known) does not exist or is unable or unwilling to communicate? SETI does indeed look very much like a wild goose chase, but on the other hand, there may be incalculable rewards just around the corner! After more than 40 years of search, no believable artificial radio signals have been recorded.

What, one might ask, is the likelihood that an extraterrestrial signal will ever be received? Or, to put it another way, can we estimate the number of advanced galactic civilizations that exist out there? Astronomer Frank Drake has established an equation that allows such a crude estimate. The Drake equation is written as follows:

$$N = N^* f_p n_e f_l f_i f_c f_L,$$

where N is the number of advanced technical civilizations in our galaxy—that is, the number of civilizations able to communicate with the rest of the galaxy by means of radio waves.

Astronomer Carl Sagan has attributed numerical values to Drake's factors as follows: N^* is the number of stars in our galaxy. We are reasonably certain that this number is 4×10^{11}. The fraction of stars with planetary systems is represented by f_p. It seems that planets are fairly common, and we can posit that f_p is $\frac{1}{3}$—that is, one third of the stars are accompanied by planets.

It is more difficult to give a value to n_e, the number of planets in a given system that are suitable for life. In our own system, we know for sure that Earth fills the requirement. However, the ecological conditions on Mars do not preclude the emergence of life. Let us say, then, that n_e equals 2.

The fraction of ecologically suitable planets on which life actually arises is represented by f_l. It is very difficult to attribute a numerical value to this factor, since we only know for sure that life exists on a single planet. On the other hand, we have seen that life is a phenomenon that derives directly from the laws of chemistry. Therefore given proper ecological conditions, such as the presence of water, a temperature compatible with life, and a significant atmosphere, Sagan estimates that f_l may be as high as $\frac{1}{3}$.

Estimating f_i and f_c proves even more difficult; f_i is the fraction of life-bearing planets on which intelligence arises, and f_c is the fraction of planets bearing intelligent life where a technical civilization also arises. Sagan estimates that

$f_i \times f_c$ is approximately $\frac{1}{100}$, meaning that only 1 percent of life-bearing planets develop technical civilizations. This is, of course, a guess; we actually have no idea what the chances are for cyanobacteria, for example, to evolve into human beings. This did take place on Earth, but we have no way to know whether it would also occur elsewhere. As we can see, N depends on a large number of factors, some known, some very poorly known, and some not much better than guesses.

At this point, multiplying all these factors together, we calculate that $N = 10^9$, meaning that the galaxy could be host to 1 billion planets harboring technical civilizations. But then, we still have to estimate f_L, the fraction of a planetary lifetime during which a technical civilization exists. For Earth, a technical civilization able to communicate with the rest of the galaxy via radiotelescopes has existed for only a few decades. Since Earth has existed for over 4 billion years, f_L for us is only about 10^{-8}. Introducing this value of f_L into the equation, we calculate that N is equal to about 10. Therefore the whole galaxy may presently contain about ten civilizations with the capacity to communicate with us, if we consider that a typical technical civilization is as old (or as young) as ours. However, our civilization may not be typical. It is quite possible that older planetary systems are hosts to technical civilizations that are millions of years more ancient than ours. Provided that these civilizations still use electromagnetic waves in the radio/television frequency range, we might be able to detect them, given enough time.

Interestingly, f_L also depends on the longevity of technical civilizations. As we know, many nontechnical civilizations have come and gone on Earth. The Babylonian, Roman, and Mongol empires, for example, all had a finite lifetime. The question then is, how stable are technical civilizations? At this juncture, one can adopt either an optimistic or a pessimistic standpoint. The optimist thinks that technical civilizations can last for a very long time, perhaps millions of years, because they eventually reach a degree of maturity that ensures long-lasting stability. If this is the case, N could be very much larger than 10. The pessimist (who is perhaps a realist) thinks that even technical civilizations can disappear in the span of a few hundred years. The cause is well known to us: war. But how could a war exterminate an entire civilization, and when might this be expected to take place?

In 1960, the British scientist L. F. Richardson compiled a formula that predicts both the magnitude of such a war and the intervals at which such wars occur. He used battlefield statistics for wars that occurred between 1820 and

1945. His formula predicts that wars killing about 1000 people should occur with intervals of about 1 month. Major conflicts killing tens of millions of people require longer intervals—about 50 to 100 years. By extrapolating into the future, Richardson's formula shows that a war claiming the entire Earth's population should take place about 1000 years from now.

The frightening thing is that we *already* possess the ability to wipe ourselves out many times over with nuclear weapons. Whether we will use them is of course entirely under our control. Thus not only does f_L in Drake's equation depend on the age of a civilization relative to the age of its planet, it also depends on the willingness of a society to not self-destruct. Therefore if and only if wars do not exist elsewhere in the galaxy, N could be as large as several million. If, on the other hand, wars are commonplace, it may well be that $N = 1$, and perhaps N will come to equal zero within the next 1000 years. We must assume that researchers of the SETI program are optimists.

These last considerations conclude what we know about the origin of the universe, and about the origins of life and its possible existence elsewhere in the solar system and in our galaxy. In this book, I have taken a strictly materialistic approach to these two problems, the only one that a scientist can seriously consider. I do believe, however, that materialistic science is not devoid of poetry and mystery. That our imagination drives us to the confines of the solar system to explore new worlds—and perhaps find evidence for extraterrestrial life or its ingredients—is poetic, even romantic. We are also a problem-solving species, and that is the reason we invented science. Once this invention was made, no stone would ever be left unturned, not even the one under which the key to the universe is hidden. But not everybody sees it this way; some people are disturbed that the science of the origins and of the evolution of life treats human beings, prokaryotes, and RNA molecules equally neutrally.

DISCONTENT WITH ORIGINS MODELS

Many people dislike the idea that science accords life, and in fact the whole universe, to be the result of totally random processes, driven only by the laws of physics and chemistry. Nature, then, and human beings too, have no purpose, no particular function in the great realm of things. Attributing a purpose to human beings seems to me, however, to be a very self-centered way of thinking. There is no evidence that we are at the culmination of an evolutionary chain, and, in spite of our efforts to modify the effects of natural selection—through medicine, for example—we are nonetheless still at the mercy of natural and self-imposed events. Sooner or later, a large comet or asteroid is bound to collide with Earth and wreak havoc on the population. We humans are profoundly altering the biosphere, thereby creating new evolutionary forces. Weapons of mass destruction, since they exist, will be used again in wars to reduce the gene pools of all terrestrial life-forms.

If indeed we have a purpose, what is it? I prefer to think that we just *are* and that our consciousness, the result of an evolutionary process that started 3.5 billion years ago, has produced one major realization in humans: we are the first and only species on Earth able to reflect on the origin and fate of the universe. Not only that, we are also able to build rational models to investigate what this origin and this future might be.

It has been a historical trend that, over the millennia, the place of Earth and that of humans has gone from full centrality to less and less preponderance in the grand scheme of things. At first, people believed that Earth was immobile in the center of the universe and all planets plus the Sun revolved around it. The distant stars were neatly arranged on a spherical surface that surrounded the Earth system. Then, Earth lost its central position in the solar system—it became a planet like many others. Later, even the Sun lost importance when it was realized that it was just another star. The discovery that the Milky Way is a vast system of many stars led to the notion that our galaxy constituted the whole universe. Bad luck! Billions of galaxies populate the universe. In short, planet Earth is one microscopically small chunk of rock, air, and water in the immensity of the cosmos. Most people have come to accept that.

But now, scientists are telling us that life appeared on Earth perhaps as a result of lightning passing through some gases. What is more, cats, dogs, and humans now derive from the coupling of bacterial and archaeal cells or, even worse, a virus! As much as the sanctity of Earth was dethroned by science in the 1600s, the sanctity of people is now also under attack. Or is it simply that there never was anything to attack? Is it perhaps simply that science is now able to tackle the origins of life, whereas it could not reasonably do so a hundred years ago? Why not just accept the results of our experiments and speculations as proof that we are an intelligent species? And besides, the experimental science of the origins of life is younger than the writer of this book. There is still much ground to cover.

As we see in this book, the results of the reflections on origins at large are often counterintuitive, tentative, speculative, and incomplete. This is not to say they are all wrong. On the other hand, one can predict that, based on previous experience, new discoveries will invariably either support or refute many of our explanations and may even provide new ones. Even though some authors have announced that the end of science is nigh, I do not think this is the case at all. Imprudent scientists have over the years declared that we knew practically all that was knowable, and they have always been proven wrong. The fun continues!

Some, however, consider that a quest for the origins of life can never be entirely successful. Among them was the late philosopher Karl Popper, who invented the falsification concept I explained in the introduction. According to him, recreating all steps that led to the appearance of life in the laboratory and even discovering life-forms or their precursors elsewhere in the universe can *never* prove that this is the way it happened on Earth. There is logic in this argument because the search for the origins of life (and of the universe) is a historical science. Barring time travel, and in the absence of witnesses, we can never be entirely certain of what happened in the distant past. However, pushing this argument too far is counterproductive because sciences such as paleontology and even geology would be relegated to the status of futile exercises. Granted, there were no human witnesses at the moment of the Big Bang and in the putative RNA world, but tracks of past events such as the microwave background radiation and ribozymes have been left for us to observe. These events must then be explained in a broad evolutionary framework. At any rate, and Popper would have agreed with this, scientific hypotheses cannot be proved absolutely true because of the open-ended nature of science. Hypotheses can be sustained or rejected only on the basis of our current knowledge and empirical observations. Therefore experiments simulating possible prebiotic chemistry, for example, may only represent possible supports for plausible hypotheses regarding the historical events that led to the appearance of life on Earth. This is because the time frame involved (hundreds of millions of years) cannot be duplicated in the laboratory.

It is at that level, however, that scientific and religious interpretations of the origins diverge most. Ancient Greek and Roman philosophers, who had rejected divine intervention in the creation of the cosmos, did build materialistic models to explain nature. However, since they did not possess adequate technology, they could not put their theoretical models to the test. In that respect, modern science is very different and much less speculative. On the other hand, religious explanations, very much across the board, have God or gods imposing their will on a void that is then transformed into the world. Life then appears not as a consequence of the creation of the world; it is created following a deliberate act of a divinity or divinities. Figure 6.2 briefly summarizes these dif-

Materialistic Interpretations

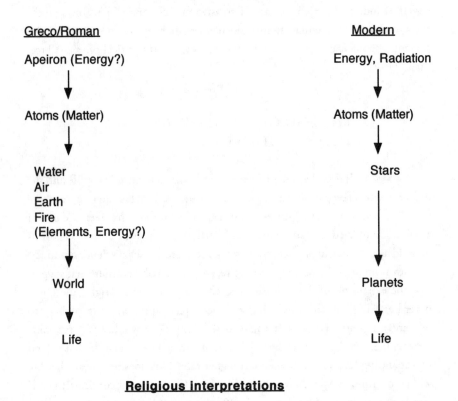

<u>Greco/Roman</u>

Apeiron (Energy?)

Atoms (Matter)

Water
Air
Earth
Fire
(Elements, Energy?)

World

Life

<u>Modern</u>

Energy, Radiation

Atoms (Matter)

Stars

Planets

Life

Religious interpretations

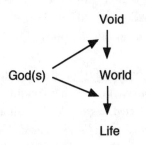

Void

God(s)

World

Life

FIGURE 6.2 A summary of three origins models. The Greco/Roman view is a composite of several hypotheses put forth by philosophers such as Thales, Anaximander, Empedocles, Democritus, and Lucretius. The modern view is a summary of the concepts described in this book. The religious interpretation is nondenominational but is shared by the world's most prevalent religions.

ferent world views. In addition, appendixes 4 and 5 summarize and briefly discuss the modern materialistic view of an astrophysicist (Eric Chaisson) and the theistic view of the neo-creationist biochemist Michael Behe. Appendix 5 also presents theological points raised by the well-known theoretical physicist Freeman Dyson.

THE END OF THE UNIVERSE,
THE END OF LIFE

It is perhaps befitting to end a book on the origins of life and the universe with a few considerations on their future. As much as no one was there to witness the birth of the universe and that of life, nobody has seen the end of a universe and lived to report about it. Nevertheless, the laws of physics allow us to build a scenario and predict how it will all end. Here is what science has to say.

Astronomers and cosmologists have recently made observations that suggest that the expansion of the universe is accelerating. If they are right, there will never be a "big crunch" in which the universe recollapses on itself, perhaps to rebound and start all over again (figure 6.3). This process would of course obliterate all life. On the contrary, we may be living in an open universe that will go on expanding forever at a higher and higher rate. This does not mean that life will go on forever. Stars have a finite life span. Our Sun will continue to convert hydrogen into helium for another 8 billion years or so. This is more than the length of time elapsed since the appearance of life on Earth. Thus barring wars and natural catastrophes, humankind (and all life-forms with it) has plenty of time left to evolve and endure.[3]

After 8 billion years, however, it will be time for our descendants to find a new home planet, either outside the solar system or perhaps on Jupiter and the other gas giants. If the gas giants are selected, a considerable amount of terraforming[4] will be necessary before an exodus can be considered. This emigration will be necessary because 8 billion years from now, the Sun will become a red supergiant and will expand until its outer layers reach the orbit of Mars (figure 6.4). Earth will be vaporized in this process. But even a massive human displacement before that happened would be only a temporary solution. After a few hundred million years in the red supergiant phase of its life, the Sun will become a cool white dwarf and eventually turn itself off. By then, whoever inhabits what is left of the solar system (including the terraformed gas giants) should have left forever.

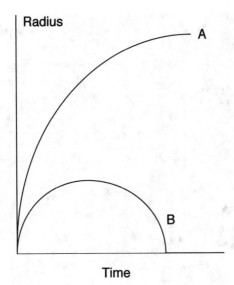

FIGURE 6.3 Two scenarios for the fate of the universe according to the theory of general relativity. **A:** The universe continues to expand forever (its radius increases infinitely). **B:** The universe collapses on itself after a period of expansion. Which scenario prevails depends on the amount of mass in the universe. We know now that there is not enough mass to produce outcome B.

But that is not all. It can be calculated that all stars will stop shining 100 trillion (10^{14}) years from now because all the fuel in the universe will be exhausted. Before that happens, due to the accelerated expansion of the universe, space will look emptier and emptier. Intelligent beings inhabiting the Milky Way and able to observe the sky with telescopes will see fewer and fewer galaxies until only Andromeda and a few others belonging to our local group remain visible. And for the grand finale, black holes will consume all galaxies by 10^{30} years from now. The universe as we know it will have ceased to exist, together with the most dedicated survivalists, if any are still there.

CONCLUSIONS

The universe and life had a beginning and they will have an end. Modern science has painted a beautiful portrait, a Vermeer or a Watteau, perhaps, of the universe and its evolution. It also tells us that our star will not go

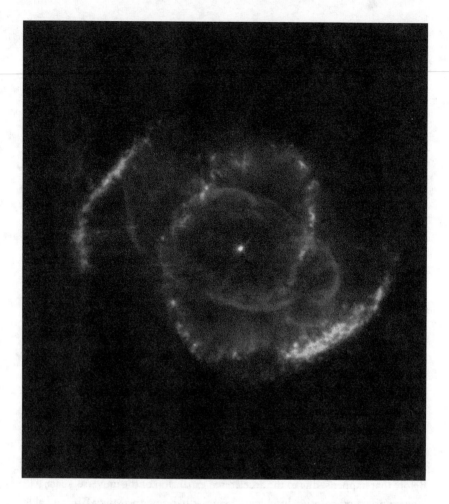

FIGURE 6.4 The Cat's Eye nebula in the Milky Way. In the course of its evolution, the central star (*dot*) has ejected much of its mass into space. Our Sun will undergo the same process a few billion years from now, extinguishing all life on Earth. (Courtesy of the National Aeronautics and Space Administration.)

extinct any time soon. The picture of the origin of life is not quite as precise and could be compared to a Jackson Pollock or a Jean Dubuffet—chaotic, but pleasing to the eye. But like a Pollock or a Dubuffet, this picture speaks to the imagination and, for some, yes, has plenty of room for improvement. We are not limited by time, since we have another 8 billion years of comfortable sunshine left to solve the riddle. If we have not done so by then, our chances are less certain. For a species that has already sent four spacecraft (the Pioneer and Voyager probes) into interstellar space, it should not take that long.

On the other hand, the end of it all need not be seen as a gloomy scenario. Before brooding over the sad ultimate fate of the solar system and that of the universe, we humans should first set our own house in order, because it may not be the thermonuclear energy of a dying, inflating Sun, or the lack thereof when all stars are dead, that will signify the end of our line. We have right now the capacity to ignite thousands of temporary Suns right here on our planet. We call them H bombs. If they are ever detonated in one of our customary wars, who cares that the Sun will stop shining a few billion years from now? The nuclear winter will have cast its dark shadow upon Earth and Earth will be lifeless.

The end of the Cold War did not cancel the threat of mutual annihilation. There are still thousands of nuclear warheads waiting to be launched. Some may be used, with unforeseen consequences. If (or when) we do use these doomsday weapons, we will never learn the origin of life, and we will have reaped the fruits of our ultimate stupidity. We will all disappear without knowing what we really are. We can do better than that.

A Graphic Representation
of Special Relativity

Time dilation and relativistic mass increase are best visualized in graphic form. First, let us remember that time varies with velocity according to the following equation (called a Lorentz transformation):

$$t' = (1 - v/c)t/\sqrt{(1 - v^2/c^2)},$$

where t' is time elapsing for an object moving at a velocity v, c is the speed of light, and t is time elapsing in a "still" reference frame from which the moving object is observed. Let us also remember that in classical Newtonian mechanics, $t' = t$, meaning that time elapses at the same rate everywhere, regardless of frame of reference and velocity. The time dilation equation shows that, in special relativity, t' depends on the ratio between the object's velocity and the speed of light. Figure A1.1 shows that at 150,000 km/s (half the speed of light), a clock ticks only about 58 percent as fast as a "still" clock. At 290,000 km/s, this rate is only about 8 percent. These velocities are not presently attainable by our spacecraft. However, time dilation can be experienced at much lower velocities, reachable by manned satellites. At 11 km/s (34,600 km/h), the equation shows that for each 100,000 seconds elapsed on Earth (t), the astronaut ages only 99,996 seconds (t') relative to us, the "motionless" Earth dwellers. After 1 year

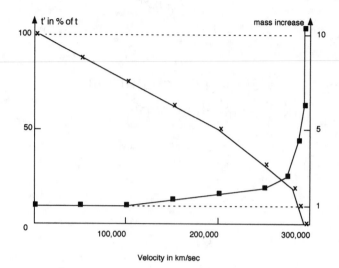

FIGURE A1.1 Time dilation and mass increase versus velocity in special relativity. *Closed squares*, mass increase as a function of velocity; *x*, the relative slowing of time as a function of velocity; *upper dashed line*, time in classical (Newtonian) mechanics (a constant as a function velocity). The *lower dashed line* represents mass in classical mechanics, which is also invariant with velocity.

in space at that velocity, the astronaut returns home 21 minutes younger than the rest of us.

In special relativity, mass varies with velocity according to the following equation:

$$m = \frac{m_0}{\sqrt{(1 - v^2/c^2)}},$$

where *m* is the mass of an object moving at a velocity *v*, *c* is the speed of light, and m_0 is the value of the object's mass at rest. Again, mass depends on the ratio between the object's velocity and the speed of light. In classical mechanics, *m* = m_0 regardless of the object's velocity. Figure A1.1 shows that a relativistic mass increase can be observed only at very high velocities. For example, at 100,000 km/s, mass increases by only about 10 percent relative to rest mass. At 290,000 km/s, *m* is about 5.5-fold the value of m_0. The shape of the curve shows that at the speed of light (300,000 km/s), the object's mass would be infinite, which is an impossibility.

More on Heisenberg's
Uncertainty Principle

The uncertainty principle is a direct mathematical consequence of the complicated equations of quantum mechanics. Werner Heisenberg did not simply postulate it;
he discovered his principle after studying these equations and pushing them to
their limit. In its energy-time formulation, the uncertainty principle states that
$\Delta E \Delta t \geq h/2\pi$, where h is the Planck constant. Thus quantum mechanics allows
for the existence of two types of subatomic particles: particles that last as long
as their Δt or longer are called *real*, whereas those that last less than their Δt are
called *virtual*. All particles can be virtual, including the familiar electron and
proton and the less familiar pions, muons, and so on. Lighter particles last
longer as virtual particles than more massive ones. This is because of Einstein's
equation $E = mc^2$. The uncertainty principle can thus be rewritten as $\Delta t \geq$
$(h/2\pi)/mc^2$. A particle whose lifetime obeys the latter equation is called real. If,
however, the lifetime of a particle is less than that prescribed by the equation,
this particle is virtual. An example will clarify this.

The mass of the proton is about 10^{-27} kg. The value of the Planck constant
is 6.62×10^{-34} joules/hertz. The value of 2π is about 6.28. The speed of light is
about 300,000 km/sec. Therefore to be real, a proton must have a lifetime of at
least the value calculated in the following equation:

$$\Delta t = (6.62 \cdot 10^{-34}/6.28)/10^{-27} \times 9 \cdot 10^{16} = 1.05 \cdot 10^{-34}/9 \cdot 10^{-11}$$
$$= 0.11 \cdot 10^{-23} \text{ sec.}$$

A proton that lasts less than this time is called a virtual proton. For an electron, whose mass is 1/1838 that of the proton, $\Delta t = 0.11 \cdot 10^{-23} \times 1838 = 202.10^{-23}$ sec. In other words, an electron can spend a much longer time as a virtual particle than a proton can, because it is much less massive. Note, however, that the time spans involved are incredibly short, even for the electron.

Virtual particles are not directly observable. However, they are responsible for the Casimir effect described in chapter 1 and other quantum effects routinely observed in particle accelerators. In other words, quantum weirdness, in addition to providing us with "smeared out" matter particles, thus also extends to the principle that "what you cannot directly see may still exist and probably does." Einstein, one of the forefathers of quantum mechanics, could never accept the idea that the subatomic world is not deterministic. So far, however, he has been proven wrong in this area. Quantum mechanics works. On the other hand, it is legitimate to ask the question, where (or at what scale) does the quantum world stop and where does our everyday macroscopic, undeniably deterministic world begin? Researchers are attempting to answer this question. There are no answers yet.

How Do We Know the Age
of the Universe?

The age of the universe is known to be 12 to 15 billion years because, ever since the pioneering work of the American astronomers Vesto Slipher, Edwin Hubble, and Milton Humason in the first three decades of the twentieth century, the universe's rate of expansion has been known. Measurements of galactic redshifts show that the farther away a galaxy is from us, the faster it recedes. In other words, redshift is proportional to distance. This is shown in figure A3.1, which graphs data collected by Hubble and Humason in 1931. Clearly, the relationship between distance and velocity of recession is a straight line. Therefore the equation representing this straight line is $v = Hd$, where v is the velocity of galactic recession, H is the slope of the line (now called the Hubble constant), and d is the distance between the observed galaxies and us. Velocity is expressed in kilometers per second, and d can be expressed in kilometers as well (although light-years or megaparsecs—1 megaparsec equals 3.26 million light-years—are more commonly used).

Thus H, the Hubble constant, is expressed in kilometers times seconds^{-1} times kilometers^{-1} (or megaparsecs^{-1}), which can be reduced to seconds^{-1}, which is of course the inverse of a time. Therefore the inverse of the Hubble

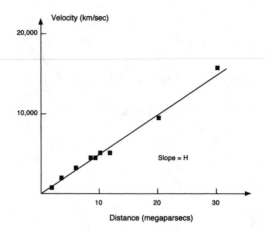

FIGURE A3.1 The relationship between galactic recession velocity (as measured by the magnitude of the redshift) and distance. The slope of the line is Hubble's constant, H.

constant, $1/H$, gives the age of the universe in seconds—that is, the time elapsed between the Big Bang and now. Current estimates of the Hubble constant range between 50 and 100 km s^{-1} megaparsec^{-1}, with an apparent consensus at 80 km s^{-1} megaparsec^{-1}. The precise value of H is not known because of the difficulties and errors encountered when measuring galactic distances.

Eric Chaisson's View
of Cosmic Evolution

I n his book *Cosmic Evolution*, astrophysicist Eric Chaisson
proposes a grand unified theory (he himself does not call it
that) of the appearance of structure, including life, in the
universe. Chaisson uses the thermodynamic theory of open systems far from
equilibrium to demonstrate that free energy rate density (ϕ_m) can be used as a
common thread to explain the increase of complexity experienced by the universe since the Big Bang.

We saw in note 1 to this book's introduction that living organisms are open
systems exchanging energy with the outside world. They are also far from equilibrium because for living systems, equilibrium means death through arrest of
metabolic functions. Chaisson extends the notion of systems in nonequilibrium
to galaxies, stars, planets, and even societies. He does indeed show compellingly
that ϕ_m increases dramatically with increasing complexity.

So, what is ϕ_m? ϕ_m is the rate of change (expressed in seconds) in free energy
(expressed in ergs, the energy unit) per unit mass (expressed in grams) experienced by a system evolving with time. Free energy, F, is a thermodynamic quantity defined by the equation

$$\delta E = \delta F + T\delta S + S\delta T,$$

where δ means change, E is the total energy of the system, S is entropy, and T is the temperature of the system. In plain English, F is energy available to do work. It can be demonstrated that in nonequilibrium open systems, structure (order) can appear locally without violating the second law of thermodynamics as long as entropy (disorder) increases in the milieu with which the system interacts (outer space in the case of galaxies, stars, and planets, and the biosphere in the case of living organisms). This allows the use of free energy to create structure.

Chaisson demonstrates that F (or rather ϕ_m) links all evolutionary steps occurring in the universe as a whole, from the conversion of radiation into matter, to the appearance of life from matter, and all the way to the appearance and development of civilization. For example, Chaisson calculates that the value of ϕ_m increases with increasing complexity and time elapsed since the Big Bang as follows:

Galaxy	0.5
Star	2
Planet	75
Plants	900
Brain	150,000
Modern society	500,000

These numbers show that free energy is a cosmic source (even perhaps a law of nature) that can be tapped to create order where only disorder existed. *Cosmic Evolution* is a fascinating but difficult book. Its level of sophistication and scientific rigor vastly outperform crude works by "scientific creationists" that contend that evolution is "just a theory" (see appendix 5).

Do the Universe and Life Have a Purpose and a Designer?

I n this book I have used a fully materialistic standpoint to explain what science knows about the origin of the cosmos and of life. This approach stems not only from my own philosophical bent but also from my belief that science and religion should not mix. I think this is an attitude espoused by an enormous majority of scientists. In this view, whether the cosmos and life have a purpose becomes largely irrelevant to the scientist searching for knowledge; we are here to study nature with our brains and our scientific tools, not to decide whether God exists. Once in a while, however, some scientists and other thinkers have crossed the science-religion barrier, usually to defend the notion that nature itself suggests the presence of a deity.

One of these thinkers was William Paley, a nineteenth-century opponent of Darwin, well known for his watchmaker metaphor. For Paley a complex object like a watch cannot appear spontaneously—it implies a watchmaker, a designer. And so it goes with life itself, said Paley. This type of argument has recently been revived by neo-creationists—including the American biochemist Michael Behe—who see in living things evidence for the existence of a creator. In general, these individuals are opposed to the theory of evolution in a Darwinian sense, although some have come up with the concept of microevolution, an ill-defined mechanism that allows for some genetic flexibility leading to minor evolutionary changes. Like Paley, scientific creationists rely heavily on metaphors, this time invoking "perfect" metabolic pathways and other cellular functions that could not have evolved from less perfect ones. Life must then have been designed. To use one of their metaphors, they say that a perfect bicycle, for example, cannot evolve from metal and rubber or from an imperfect bicycle without the intervention of a designer. Since the designer of a bicycle has a purpose in mind—the transportation of people—and since life was also designed, life (and, by extension, the whole universe) must also have a purpose.

There are serious problems with this type of thinking. First, metabolic pathways are not necessarily perfect. For example, the enzyme ribulose bisphosphate carboxylase/oxygenase—the most abundant protein on planet Earth because it is present in all plants—converts atmospheric carbon dioxide into organic carbon, which is then used to make sugars. But the oxygenase function of this enzyme uses atmospheric oxygen to degrade a portion of the sugars it helps make. These sugars—and the energy consumed by plants to make them—are thus wasted for further metabolic processes. Clearly, this enzyme is quite imperfect.

Another argument used by the defenders of the designer hypothesis is the existence of the "perfect" (and very complex) blood-clotting mechanism present in vertebrates, which they say could not have evolved from a less perfect pathway. But now that large parts of the human genome have been sequenced, we know that the human blood-clotting system *is* present in simpler form in the fruit fly and even in a worm, two invertebrates. Clearly, the complex system found in humans and other vertebrates results from the recruitment of more ancient genes (such as worm and fly genes) often accompanied by their rearrangement. This observation demonstrates that the vertebrate blood-clotting mechanism did in fact evolve from a simpler system and was not designed with a specific purpose in mind. On the contrary, organisms that acquired a more efficient blood-clotting mechanism through evolutionary processes recovered more quickly from injury, and they lived and reproduced more successfully, thereby passing on their better adapted genes.

Many more examples exist to show that cellular functions are far from perfect and must have developed from ancestral forms. However, as has happened so many times in history, modern science cannot prove any better than before the existence of a divine designer. As the famous theoretical physicist—and devout Christian—Freeman Dyson has written in relation to the inflamed debates between Thomas Huxley, the defender of Darwin, and Bishop Wilberforce, his vitriolic opponent and champion of divine design, "Looking back on the battle a century later, we can see that Darwin and Huxley were right."

Dyson is an extremely interesting individual. With Richard Feynman, the American physicist and Nobel laureate, he contributed in the 1950s to the development of quantum electrodynamics, a theory that some call the most verified in all of physics. An Englishman, he became a professor at the Institute for Advanced Studies in Princeton, New Jersey, where he soon turned his attention to futuristic aspects of scientifically advanced civilizations. For example, he invented the concept of the Dyson sphere, an enormous hollow contraption with the dimensions of a solar system, designed to capture practically all the energy emitted by a star. This sphere could be constructed by mining planets orbiting the star, it would be inhabited on its inner surface (providing immensely great space for an exploding population), and it would never need an energy source other than its star. This concept has been used, with modification, in the intriguing science fiction novel *Ringworld* by Larry Niven.

Interestingly, Dyson is one of the few scientists I know of who explicitly states that the cosmos and life do not make sense without the existence of God. However, as I said earlier, Dyson does not espouse the idea that God is some type of engineer who mapped out the details of the Big Bang and designed living organisms to make them perfect. Dyson's God is far more subtle. For Dyson, the universe cannot be an accident—that is, a chance event. Indeed, if the masses of elementary particles had been created very different from the existing ones, much of physics and chemistry would have been different, and life as we know it might not have appeared. Similarly, if the four fundamental forces were very different in their relative strengths (for example, if gravity superseded the other three forces), the universe would be a very different place today. Thus Dyson sees in the laws of physics not a proof of the existence of God but an indication that the "architecture of the universe is consistent with the hypothesis that mind plays an essential role in its functioning." It should be kept in mind that for Dyson, apparently, mind and soul are a single concept. And, indeed, without the presence of a mind (or minds or souls) that tries to understand the universe and discover its laws, the universe might as well not exist. However, Dyson makes it very clear that this is as far as science can go. For him the existence of a world soul (God) is a question that belongs to religion and not to science. Dyson the Christian does believe in a world soul, however, and has thus gone beyond the threshold that Dyson the physicist could not cross.

Interestingly, Dyson's 20-year-old argument has been resurrected by some modern cosmologists in the form of an "anthropic principle." This principle states that of all the possible universes that could have emerged from the Big Bang, only ours was habitable, as we are here to prove it! This seems to be yet another teleological (and circular) argument, particularly hard to falsify, since these "other" universes are not observable. In effect, there may be no difference between an anthropic universe and a universal mind or soul à la Dyson.

In contrast to these world views, the late French biologist Jacques Monod takes a strict materialistic position. For Monod, "The cornerstone of the scientific method is the postulate that nature is objective. In other words, the *systematic* denial that true knowledge can be got at by interpretating phenomena in terms of final causes, that is to say, of purpose." This means that for Monod (as for Weinberg), the universe has no purpose and any teleological implication of its creation must be rejected. Note that Monod *postulates* that nature is objective, whereas, in a sense, Dyson *postulates* that the laws of physics are not accidental. These positions are irreconcilable, and in the final analysis, at least in the realm of science, a Dysonian postulate cannot be maintained, as he himself has recognized. If indeed nature were not objective, we scientists would have to pack our bags immediately and go home, as there would be nothing left for us to do. This is not to say that such deep and difficult questions should not be raised. For us humans issues of relevance, free will, and metaphysics will always exist. Some will think they know the answers, and others will disagree. Forever.

Notes

INTRODUCTION

1. Some people claim that life must have had a supernatural origin because it violates the second law of thermodynamics. Simply put, the second law dictates that all natural systems spontaneously tend toward a state of thermodynamical equilibrium characterized by maximum disorder. One example is a cube of sugar in a cup of hot coffee. When you drop the lump of sugar into your coffee, for a fleeting instant, the lump remains intact and the coffee remains unsweetened. This is an initial state of order: the sugar remains solid and the coffee remains bitter. As time goes by, the sugar dissolves and totally loses its cubic nature. Sugar molecules and molecules contained in the coffee become thoroughly mixed. This is the final state of total disorder. No matter how long you wait, your sweetened coffee will not spontaneously generate unsweetened coffee and an intact sugar cube. In a different type of formulation, the second law says that entropy of an isolated system must increase, entropy being a measure of disorder. The key term in the formulation of this law is the word *isolated*. An isolated system is one that does not exchange energy or matter with the outside world. A living system is clearly not an isolated system: it receives food (matter) from the outside world and also energy (as food is digested and oxygen is breathed and this process used to synthesize energy-rich compounds). Even the whole planet Earth is not an isolated system; it receives enormous amounts of energy in the form of electromagnetic radiation from the Sun. In short, living systems do not violate the second law because they are not isolated systems. Erwin Schrödinger explains this very well in his little book (see the bibliography).

2. Some opponents to evolution have likened the all-at-once creation of a living cell to putting scrap metal in a sack, shaking it well and coming up with a Rolex watch. I agree with them. This example shows that living cells were not created all at once but that their functions and structure evolved gradually, from simple to more complicated, through natural selection.

   Many people have difficulties with this concept. The following example should help. Consider a system in which two functions (such as chemical reactions) compete with one another for the same compound, called the substrate. Let us assume that the medium in which the reactions are taking place contains two different "facilitators" that drive these two reactions. One facilitator is fast and the other is slow. The fast facilitator makes the conversion of the substrate into another product efficient and fast, while the slow facilitator does so only very slowly. As the fast reaction depletes the substrate, the slow one is left with nothing to sink its teeth into and ceases to operate. One can say that this slow reaction is not adapted to the prevailing conditions. If the substrate used in the reaction is plentiful, the fast reaction overtakes the slow one. On the other hand, if the substrate becomes depleted, and if the slow reaction is able to use a slight variant of this substrate (assumed to be present) while the fast one cannot, then it is the fast reaction that ceases to operate and the slow one takes over. Thus depending on conditions, either type of reaction may be more successful than the other.

It is not difficult to see how multiple reactions, some competing, some cooperating, can lead to fast evolution of a system constrained by prevailing external conditions, such as compound availability, temperature, and salinity, that can all potentially affect reaction rates. In the end, these external conditions select the reactions that become dominant and those that play a very secondary role in the system and eventually disappear.

3. Democritus's motto was echoed by Jacques Monod. In his book *Chance and Necessity*, Monod offers fascinating thoughts on the nature of biology and its philosophical implications. This book has been criticized (mostly in the Anglo-Saxon world, but not in France), unjustly I believe, for its gloomy, postexistentialist view of human nature. One of Monod's concluding remarks, "The ancient covenant is in pieces; man knows at last that he is alone in the universe's unfeeling immensity, out of which he emerged only by chance," has been widely misinterpreted. In my opinion, Monod was referring to the *existential* loneliness of man (today we would say "humans"), not to the fact that life and even intelligence could not exist elsewhere in the universe. Since life on Earth was created by chance, and evolution led to us, humans, by necessity, it would be extraordinarily unlikely that our carbon copies would exist elsewhere in the universe. Hence, our loneliness. This does not preclude extraterrestrial life and extraterrestrial intelligence. What is more, Monod prefaces his book with an extraordinary quote from Albert Camus's "The Myth of Sisyphus." Sisyphus, the son of Aeolus, the wind god, has been punished by the gods for various crimes. His punishment for eternity is to roll a huge stone to the top of a hill only to have it roll back down as it reaches the top. This myth is classically seen as a symbol of futility. However, Camus radically revises this interpretation and concludes his essay with the sentences, "The struggle itself toward the heights is enough to fill a man's heart. One must imagine Sisyphus happy." These must be two of the most optimistic phrases ever written. Monod clearly abides by this statement. He died prematurely of leukemia 25 years ago, a happy individual, I am sure.

CHAPTER 1

1. The notion of a still or motionless object can be understood only in terms of reference frame. For example, you and I can sit still in the frame of reference that is our room or even planet Earth. However, we are not truly sitting still because, for a reference frame located away from Earth (on Mars, for example), Earth is rotating and we are rotating with it. In addition, Earth moves in orbit around the Sun. Thus if the Sun is used as the reference frame, you and I are definitely not motionless. The Sun, in turn, is in orbit around the galactic center while galaxies of our local group orbit one another. The local group of galaxies is also in motion, and so on. In other words, there is no such thing as an absolute reference frame.

2. Not all galaxies are receding from our own. Our neighbor, the Andromeda galaxy, is actually rushing toward the Milky Way at 300 km/sec and its light is accordingly blueshifted. This is because the Milky Way and Andromeda, together with another thirty smaller galaxies, are part of a group whose members are gravitationally bound to each other. Galaxies are often organized in clusters and the members of a cluster may or may not recede from one another. However, whole clusters do recede with the expansion of the universe.

3. You car does not seem to move by successive jerks (unless the gas line is plugged up) because at the macroscopic level, discontinuous energy distribution resulting from quantum effects is not noticeable. Quantum effects are smoothed out because of the sheer number of individual events taking place in your motor.

4. It is ironic that the model of the atom proven wrong 75 years ago is still depicted on T-shirts and company logos. It is also taught in high school. This model shows the atom as a miniature solar system, where the nucleus is a Sun of sorts and the electrons are planets orbiting that star. This model is outmoded, but it is so simple that it continues to be used.

5. By *ordinary matter*, I mean matter that you and I experience on a dayly basis. That matter is made of protons, neutrons, and electrons. *Extraordinary matter* does exist in the universe. Odd particles are found in cosmic rays and high-energy collisions in particle accelerators. All these particles, however, are short-lived and can be understood in terms of the "standard model" of physics. They too are composed of quarks and they have their own accompanying leptons. These particles have no significance in ordinary affairs (including life) on Earth today.

In addition to this extraordinary matter, the possible existence of mysterious "dark" matter must also be considered. Dark matter was first invoked to explain why different parts of galaxies rotate around their center at roughly the same velocities. Furthermore, it has been calculated that clusters of galaxies could not remain stable without the presence of a large amount of invisible (nonluminous) matter. It is now estimated that up to 90 percent of the universe's matter is dark matter. We know that extinguished stars or neutrinos cannot possibly account for this figure. In other words, dark matter is not made of atoms or familiar subatomic particles. No one knows the nature of dark matter, but it may consist of particles without electric charge that interact very weakly with ordinary matter. One of the biggest challenges facing modern cosmology is to discover what dark matter is and how it was formed.

CHAPTER 2

1. The concept of inflation also helps explain the "flatness" of the universe. Flatness here refers to the curvature of space-time, which can be flat, open, or closed. A flat universe is one in which there is enough matter to slow down expansion but not quite enough to halt it. Until recently, the universe seemed to be of the flat type. Recent observations, however, suggest that rather than decelerating, expansion is accelerating. If these results are confirmed, we are living in an open universe. At any rate, a collapse of the universe, which would happen if it were closed, does not presently seem plausible.

2. The Planck time is defined by the three fundamental constants G (Newton's gravitational constant), h (the Planck constant), and c (Einstein's speed of light). The units in which these constants are expressed cancel each other out, except time. Thus the Planck time is represented by the following equation:

$$t_P = \sqrt{(h/2\pi)G/c^5} = 5.39 \cdot 10^{-44} \text{ second.}$$

The Planck length is $1.62 \cdot 10^{-35}$ meter, and it is the length crossed by light within the Planck time. Smooth space-time does not exist below the Planck time and the Planck length because of quantum fluctuations.

3. Quantum fluctuations of the vacuum produce an equal number of matter and antimatter particles. When these particles meet, they annihilate to produce photons. For reasons of symmetry, the Big Bang is expected to produce equal amounts of matter and antimatter and the universe should consist of space filled with photons. We know this is not the case. Further, matter vastly dominates antimatter in today's universe. This means that more matter than antimatter appeared shortly after the Big Bang. Why symmetry was broken is not known for sure. A mechanism involving a very massive X particle (up to 10^{16} times more massive than a proton) may have slightly favored the transformation of antimatter into matter over the reverse, explaining the existence of matter in the universe. Other examples of symmetry violation have been discovered in nuclear physics. There exist today in the universe 1 billion photons for each matter particle. These photons were created in the early universe, as quarks and antiquarks annihilated. Excess quarks surviving annihilation went on to form matter as we know it.

4. Gravitational waves are not expected to have been obscured by the plasma that pervaded the universe until it was 400,000 years old. However, these waves have not yet been detected. Two large facilities to detect gravitational waves are being built in Washington State and Louisiana and should soon be operational. They might shed light on what happened in the universe during the inflationary period.

CHAPTER 3

1. Some bacterial, plant, and animal viruses possess an RNA genome. One example is the human immunodeficiency virus (HIV), which causes AIDS in humans. This does not contradict the statement that all life is DNA-based. First, viruses are not truly alive, in the sense that they cannot manufacture their own constituents all by themselves. They need to infect a host to do that. Furthermore, they cannot independently use energy from the outside world, as living cells do. Finally, viruses are not *cells* (even those whose genome is made of DNA); they are simply RNA or DNA wrapped up in a protein coat that may or may not also contain lipids. All cellular life is invariably based on DNA. Since cellular DNA is expressed via an RNA intermediate, RNA-containing viruses skip a step in the transfer of genetic information.

2. The nature of catalysts in general and enzymes in particular should be correctly understood. The catalytic converter in your car and enzymes in living cells do not make impossible chemical reactions possible. Scientists can determine which chemical reactions are possible (and which are impossible) by thermodynamic calculations. Thermodynamics is a branch of physical chemistry that allows us to calculate the parameters of a system such as free energy and entropy. Many possible chemical reactions do not occur spontaneously, and many others occur very slowly. This is because of the existence of an activation energy barrier that must be overcome for the reaction to take place. The activation energy is that which is necessary for molecules to collide in such a way that they will react with one another. The role of a catalyst is to lower this activation energy barrier. Enzymes lower the barrier by holding the reactants in very close proximity, in "pockets" tailored for just these reactants. Enzymes thus considerably speed up chemical reactions that are thermodynamically possible.

3. Not all mutations are caused by mistakes made spontaneously by DNA polymerase. The structure of DNA can be altered by chemicals, called mutagens, that can induce much higher rates of replication errors. Particularly nasty mutagens are dioxin and benzopyrene, the latter found in tobacco smoke. Ultraviolet light and X rays are also mutagenic.

CHAPTER 4

1. Very interestingly, some meteorites also contain amino acids and other organic compounds made in Miller's experiments. This adds much credence to the fact that these molecules can indeed be synthesized in places other than Earth's surface under abiotic conditions. However, the mechanisms of formation of these molecules in meteorites are not known. One famous meteorite containing organic compounds is the Murchison meteorite found in Australia. The types of compounds in this meteorite as well as their relative amounts are remarkably similar to those found in Miller's experiment.

2. For example, HCOOH is formic acid (produced by some species of ants) and CH_3COOH is acetic acid—that is, vinegar.

3. I use the term *nerd* without any disrespect. As a scientist, I myself must have appeared as a nerd, a prima donna, and maybe even a wise man over the course of my life.

CHAPTER 5

1. Eigen's group demonstrated that the hypercycle model accounts very nicely for the life cycle of bacteriophages. The hypercycle is thus not just a fancy mathematical model.

2. Eigen and coworkers developed a complex mathematical theory, which they dubbed statistical geometry, to study the evolution of nucleic acid sequences and that of the genetic code.

3. For example, gene duplication explains very well the origin of genes that code for the protein part of hemoglobin and myoglobin, globin. Humans can produce six types of hemoglobin and one type of myoglobin, an oxygen-binding protein found in muscle. It has been shown that the corresponding seven genes derived from a single ancestral copy that duplicated over time. Globin gene analogs are even found in some plant species.

4. The hydrogen hypothesis does not explain the existence of chloroplasts. We must assume that once formed, the proto-eukaryote would have engulfed cyanobacteria to give rise to plant lineages.

CHAPTER 6

1. How can scientists tell the difference between Martian, lunar, and asteroid belt meteorites? The key lies with space exploration. Spacecraft have landed on the Moon and Mars to analyze rock and soil samples as well as gases trapped in them. Asteroid belt bodies have also been scrutinized from nearby. It has been found that the composition of meteorites carries the signature of their origin.

2. President George W. Bush publicly announced on August 11, 1994, when he was still Texas governor, that Mars has an orbit similar to that of Earth, canals where water flows, and a breathable atmosphere. Canals and a breathable atmosphere were proved nonexistent by spacecraft that orbited around and landed on Mars in 1976. That Mars and Earth do not occupy similar orbits was known 500 years ago.

3. Science never ceases to baffle me. As I was reviewing my copyedited manuscript, two cosmologists published a radically different model of the universe, a model in which the universe expands and contracts in an infinite number of cycles (see *Science*, issue number 5572, 2002, pp. 1417-1433 and 1436-1439). This new model does away with the nagging problem of the singularity at time zero. In the stan-

dard model, time has a beginning. In the new cyclic model, time has no beginning—it is simply infinite, in the past and in the future.

We have seen that general relativity predicts that an expanding universe can collapse if its curvature is positive. There would then be a Big Bang followed by a Big Crunch. A universe with positive curvature has been ruled out by observations that indicate that there is not enough matter-energy in the universe to cause this type of curvature. We now know that the geometry of the universe is flat, not curved. How, then, can one build a model of a Big Crunch?

In the new cyclic model, what causes collapse after expansion is negative potential energy, also called dark energy. This type of energy has been hypothesized to explain the apparent increase in the rate of expansion of the universe. The cyclic model predicts that after trillions of years, the rapidly expanding universe will return to a complete vacuum state devoid of matter and radiation, and it will become momentarily static. At this point, the universe starts contracting; its negative potential energy decreases until it reaches a minimum (the Big Crunch), and the universe bounces back through a Big Bang, and this is followed by the appearance of radiation and then matter, as in the standard model. This process is repeated an infinite number of times.

The cyclic model accounts for what we know about the universe—that its geometry is flat, that expansion is accelerating, that it contains dark matter, and that density fluctuations must have been produced at the Big Bang. It also solves the problem of the beginning of time and space, since it posits that there was no such beginning. When I think about this model, I cannot help recollecting dancing Shiva, as he is about to destroy the universe and re-create it in an endless number of cycles. I also imagine that a Hindu priest would be thoroughly unimpressed by the new cyclic model just generated by complicated mathematical and physical thinking. I presume his answer would be, "I always knew that and I told you so a very long time ago."

4. Terraforming consists of altering a planet's surface and its atmosphere to suit human needs. We do not yet have this technology.

Glossary

aerobic: relating to life or metabolic reactions that take place in the presence of oxygen

amino acids: the building blocks of proteins; composed of a central carbon atom linked to a hydrogen atom, a variable side group, a carboxyl group (COOH), and an amino group (NH_2)

amphiphilic: meaning "loves both" and relating to the ability of phospholipids to interact at the same time with water and hydrocarbons

anaerobic: relating to life or metabolic reactions that take place in the absence of oxygen

anticodon: a set of three contiguous bases in transfer RNA whose role it is to decode a codon

antimatter: a type of matter in which electrons are positively charged and protons negatively charged; very rare in the universe but easily produced in particle accelerators

ATP (adenosine triphosphate): an energy-rich molecule composed of adenine, ribose, and three phosphate groups all linked together

bacteriophage: a virus that infects prokaryotes

baryons: a family of heavy subatomic particles, such as protons and neutrons

base: one of the constituents of RNA and DNA. RNA contains the four bases adenine (A), guanine (G), cytosine (C), and uracil (U); in DNA, thymine (T) replaces U. (See *nucleotide*.)

blackbody: an object that absorbs and reemits thermal energy in the form of electromagnetic radiation

catalyst: a compound or an element that facilitates a chemical reaction without being modified by it (e.g., platinum in a catalytic converter)

codon: a set of three contiguous bases in DNA; it specifies an amino acid

cosmic rays: intense radiation, usually in the form of subatomic particles, that pervades the universe. The origin of the most energetic cosmic rays is poorly understood.

cosmology: the science of the universe, the cosmos

cytoskeleton: a mesh of proteins that gives eukaryotic cells some rigidity and mobility

DNA (deoxyribonucleic acid): a long, double-helical molecule containing base pairs in its center and a backbone of deoxyribose and phosphate

electromagnetic field: an oscillating field associated with moving electrical charges

electron: a negatively charged lepton surrounding an atomic nucleus

enzyme: a biological catalyst

error threshold: the frequency of replication errors beyond which a quasispecies cannot remain stable

eukaryote (also Eukarya): cells characterized by the presence of a nucleus that contains most of that cell's DNA

exon: the portions of eukaryotic genes that code for amino acids and are interrupted by introns

field: a region of space where a force exists (e.g., the gravitational field of a star)

force: a quantity relating the mass of a body to its acceleration ($F = m \times a$). Energy is force times distance.

frequency: the number of cycles per second undergone by periodic phenomena such as electromagnetic waves. A high frequency corresponds to a short wavelength, whereas a low frequency corresponds to a long wavelength.

galaxy: a vast collection of gravitationally bound stars. The Milky Way is sometimes called the Galaxy (note the capital G).

gamma rays: hard electromagnetic radiation of very high frequency

gene: a string of DNA base pairs that generally code for a protein

genotype: the suite of all genes harbored in an organism

gravitation: an attractive force exerted between two or several massive bodies

greenhouse effect: increase in temperature of a planet's surface as a result of the presence of gases such as carbon dioxide and methane that trap infrared radiation and send it back to the surface

histone: a family of proteins that are found in the nucleus of eukaryotic cells and that interact with DNA

hydrocarbon: an organic molecule consisting of carbon and hydrogen atoms

hypercycle: a group of cooperating quasispecies

intron: a DNA sequence that interrupts a gene and does not code for amino acids

inverse square law: a mathematical law that states that the effects of certain force fields decrease with the inverse of the square of the distance separating two objects

ion: an electrically charged atom or molecule having gained or lost one or several electrons. Many compounds become ionized when dissolved in water. Ions can also be formed when energy, in the form of heat, for example, is provided to atoms or molecules.

isotope: an atom that contains the same number of protons but a different number of neutrons. Isotopes of an element have the same chemical properties.

leptons: a family of light subatomic particles, such as electrons and neutrinos

lipids (also called fatty acids): long-chain hydrocarbons, which, when found in cell membranes, contain phosphate and are called phospholipids

liposome: an artificial microscopic vesicle made of a phospholipid bilayer

mass: a property of matter and radiation characterized by inertia (resistance to changes of motion)

messenger RNA (mRNA): the RNA copy of a DNA gene

metabolism: the set of chemical reactions that occur in living cells

microorganism: an organism, usually unicellular, that can be seen only with a microscope

mutation: the change of one DNA base pair into another

natural selection: the process that results in improved (or decreased) ability to survive when natural agents—such as climate and food availability—favor (or act against) a particular set of genes possessed by cells or organisms

neutrino: a lepton produced in thermonuclear reactions and certain types of radioactive decay. Neutrinos interact very weakly with matter.

neutron: an electrically neutral baryon found in atomic nuclei. Isolated neutrons are unstable and decay.

nucleosome: a eukaryotic structure in which DNA is wrapped around a protein core formed of eight histones

nucleosynthesis: the process by which atomic nuclei were formed in the young universe

nucleotide: a compound formed by an association between a base and the sugar ribose (in RNA) or deoxyribose (in DNA) and one or several phosphate groups. The length of a molecule of DNA or of RNA is given in numbers of nucleotides or bases (these numbers are the same) aligned in the molecule.

nucleus (plural, nuclei): (1) in an atom, the central portion formed of protons and neutrons (except for hydrogen, which has only a single proton in its nucleus) where practically all the mass of the atom is concentrated; (2) in eukaryotic cells, a body that contains most of the cellular DNA

particle: in physics, a body that is part of an atom, a proton, or a neutron or that is associated with a force field. Quarks and electrons, for example, are particles.

pathway: in biochemistry, a series of metabolic reactions that convert a precursor into a final product

phenotype: expression of the genotype in the form of recognizable characteristics, such as shape, color, and metabolism

photons: particles of light. Photons are particles of zero rest mass that always move at the speed of light.

photosynthesis: the mechanism by which organisms capture sunlight's energy to perform metabolic reactions

phylogeny: the study of the ascent, genetic relationships, and evolutionary history of living and extinct organisms

polymerization: the act of linking together small building blocks to make long molecules. DNA, proteins, rubber, and plastics are polymers.

prebiotic: relating to Earth or to chemical reactions as they existed in a world devoid of life

precursor: a compound destined to undergo a chemical modification to generate a final product

progenote: the ancestor of all DNA-containing cells

prokaryote: a cell in which DNA is not confined in a nucleus. For example, Archaea and Bacteria are prokaryotes.

proteins: long chains of amino acids linked together by chemical bonds

proton: a positively charged baryon found in atomic nuclei

quantum mechanics: the part of theoretical physics that studies the properties of subatomic particles by treating them as matter waves

quasispecies: a "cloud" of mutant RNA or DNA sequences deriving from a master sequence

redshift: the shifting toward longer (redder) wavelengths of the light emitted by a receding object

reduction-oxidation: two chemical reactions that result in the gain of electrons and often hydrogen atoms (reduction) or their loss (oxidation)

replication: the mechanism by which DNA copies itself into two identical daughter double helices

ribosome: a small cellular body composed of ribosomal RNA and proteins, where protein synthesis takes place

ribozyme: an RNA molecule with enzymatic, catalytic activity

RNA (ribonucleic acid): a single-stranded molecule composed of many subunits of base, ribose, and phosphate linked together

spectrum: the ensemble of all electromagnetic radiation found in nature, from radio waves (low frequency and energy) to gamma rays (high frequency and energy)

terraforming: turning the surface of an inhospitable planet into an environment suited to human habitation (not yet achieved)

thermodynamics: the science of energy and heat transactions

thermonuclear: relating to fusion reactions between two or several atomic nuclei

transcription: the mechanism by which DNA is copied into an identical base sequence made of RNA

transfer RNA (tRNA): Small RNA molecules that bind amino acids and position them in the correct place along the molecule of messenger RNA being translated

translation: the mechanism by which the RNA copy (the mRNA) of the DNA gene is decoded by tRNAs (with the help of ribosomes) to make proteins

uncertainty principle: the principle of quantum mechanics that states that the position and the velocity of a subatomic particle cannot be known at the same time with infinite accuracy

wave: a region of space where a force or matter field varies

wavelength: the distance separating two crests of a wave, such as in an electromagnetic wave

Notable Scientists

ARRHENIUS, SVANTE (1859–1927): Swedish chemist who invented the concepts of ions and panspermia

BOHR, NIELS (1885–1962): Danish physicist who developed the concepts of stationary electronic orbits and quantum jumps

BORN, MAX (1882–1970): German physicist who developed the concept that the position of an electron is determined by probability equations

DARWIN, CHARLES (1809–1882): English naturalist who originated the concept of evolution by natural selection

DE BROGLIE, LOUIS (1893–1987): French physicist who demonstrated that a particle has an associated wavelength

EINSTEIN, ALBERT (1879–1955): multinational physicist who invented the concepts of special and general relativity

GALILEO, GALILEI (1564–1642): Italian physicist and astronomer who formulated the laws of motion and discovered the satellites of Jupiter

HEISENBERG, WERNER (1901–1976): German physicist who was the co-inventor of quantum mechanics and author of the uncertainty principle

HUBBLE, EDWIN (1889–1953): American astronomer who discovered that the redshift of galaxies is proportional to their distance from a reference frame

LEMAÎTRE, GEORGES (1894–1966): Belgian mathematician who first enunciated the principle of the big bang

LORENTZ, HENDRIK (1853–1928): Dutch physicist who first formulated the equations of motion that take the speed of light into account

MAXWELL, JAMES CLERK (1831–1879): Scottish mathematician and physicist who invented the concept of electromagnetic waves that propagate at the speed of light

NEWTON, ISAAC (1642–1727): English physicist who co-invented calculus (with Gottfried Leibnitz) and formulated the law of universal gravitation

OPARIN, ALEXANDER (1894–1980): Soviet biochemist who first hypothesized that the building blocks of life could have been synthesized in a reducing atmosphere

PLANCK, MAX (1858–1947): German physicist who discovered that energy comes in finite quantities, which he called quanta. Energy is thus not infinitely divisible. The energy of electromagnetic radiation is directly proportional to its frequency.

SAGAN, CARL (1934–1996): American astronomer, controversial author, and polymath, who correctly inferred that the atmosphere of Venus creates a strong greenhouse effect.

SCHRÖDINGER, ERWIN (1887–1961): Austrian physicist and co-inventor of quantum mechanics

Bibliography

Introduction

Chaisson, E. J. 2001. *Cosmic Evolution: The Rise of Complexity in Nature.* Cambridge, Mass.: Harvard University Press.

de Duve, C. 1991. *Blueprint for a Cell: The Nature and Origin of Life.* Burlington, N.C.: Neil Patterson.

Dyson, F. 1981. *Disturbing the Universe.* New York: Harper Colophon Books.

Fry, I. 2000. *The Emergence of Life on Earth: A Historical and Scientific Overview.* New Brunswick, N.J.: Rutgers University Press.

Monod, J. 1971. *Chance and Necessity: An Essay on the Natural Philosophy of Modern Biology.* New York: Alfred A. Knopf.

Schrödinger, E. 1947. *What Is Life? The Physical Aspect of the Living Cell.* New York: Macmillan.

Chapter 1

Barrow, G. M. 1996. *Physical Chemistry.* New York: McGraw-Hill.

Emiliani, C. 1992. *Planet Earth: Cosmology, Geology, and the Evolution of Life and Environment.* Cambridge, England: Cambridge University Press.

Feynman, R. P. 1997. *Six Not-So-Easy Pieces: Einstein's Relativity, Symmetry, and Space-Time.* Reading, Mass.: Addison-Wesley.

Kahan, T. 1959. *Théories Quantiques.* Paris: Librairie Armand Colin.

Sagan, C. 1980. *Cosmos.* New York: Random House.

Spielberg, N., and B. D. Anderson. 1987. *Seven Ideas That Shook the Universe.* New York: John Wiley & Sons.

Chapter 2

Emiliani, C. 1992. *Planet Earth: Cosmology, Geology, and the Evolution of Life and Environment.* Cambridge, England: Cambridge University Press.

Gonzalez, G., D. Brownlee, and P. D. Ward. 2001. Refuges for life in a hostile universe. *Scientific American* 285:60–67.

Gribbin, J. 1998. *In Search of the Big Bang: The Life and Death of the Universe*. Harmondsworth, Middlesex, England: Penguin Books.

Hawking, S. 1990. *A Brief History of Time: From the Big Bang to Black Holes*. New York: Bantam Books.

Hogan, C. J. 1998. *The Little Book of the Big Bang: A Cosmic Primer*. New York: Springer-Verlag.

Padmanabhan, T. 1998. *After the First Three Minutes: The Story of Our Universe*. Cambridge, England: Cambridge University Press.

Rees, M. J. 2000. Piecing together the biggest puzzle of all. *Science* 290:1919–1925.

Shklovskii, I. Undated (ca. 1964). *Univers, Vie, Raison*. Moscow, USSR: Editions de la Paix.

Weinberg, S. 1993. *The First Three Minutes: A Modern View of the Origin of the Universe*. New York: Basic Books.

CHAPTER 3

Ayala, F. J. 1982. *Population and Evolutionary Genetics: A Primer*. Menlo Park, Calif.: Benjamin Cummings.

Becker, W. M., L. J. Kleinsmith, and J. Hardin. 2000. *The World of the Cell*. San Francisco: Benjamin Cummings.

de Duve, C. 1991. *Blueprint for a Cell: The Nature and Origin of Life*. Burlington, N.C.: Neil Patterson.

de Duve, C. 1995. *Vital Dust: Life as a Cosmic Imperative*. New York: Basic Books.

Hartwell, L. H., L. Hood, M. L. Goldberg, A. E. Reynolds, L. M. Silver, and R. C. Veres. 2000. *Genetics: From Genes to Genomes*. New York: McGraw-Hill.

Jacob, F. 1973. *The Logic of Life: A History of Heredity*. Princeton, N.J.: Princeton University Press.

Lurquin, P. F. 2002. *High Tech Harvest: Understanding Genetically Modified Food Plants*. Boulder, Colo.: Westview Press.

CHAPTER 4

Bernstein, M. P., S. A, Sandford, and L. J. Allamandola. 1999. Life's far-flung raw materials. *Scientific American* 281:42–49.

Brack, A. 1998. *The Molecular Origins of Life*. Cambridge, England: Cambridge University Press.

de Duve, C. 1991. *Blueprint for a Cell: The Nature and Origin of Life*. Burlington, N.C.: Neil Patterson.

de Duve, C. 1995. *Vital Dust: Life as a Cosmic Imperative*. New York: Basic Books.

Hazen, R. M. 2001. Life's rocky start. *Scientific American* 284:77–85.

Maynard Smith, J., and E. Szathmry. 1997. *The Major Transitions in Evolution*. Oxford, England: Oxford University Press.

Oparin, A. Undated (ca. 1964). *L'Origine et l'Evolution de la Vie*. Paris: Editions de la Paix.

CHAPTER 5

de Duve, C. 1991. *Blueprint for a Cell: The Nature and Origin of Life.* Burlington, N.C.: Neil Patterson.

de Duve, C. 1995. *Vital Dust: Life as a Cosmic Imperative.* New York: Basic Books.

Eigen M. 1992. *Steps Towards Life: A Perspective on Evolution.* Oxford, England: Oxford University Press.

Eigen, M., and R. Winkler-Oswatitsch. 1981. Transfer-RNA: An Early Gene? *Naturwissenschaften* 68:282–292.

Eigen, M., and R. Winkler-Oswatitsch. 1981. Transfer-RNA: The Early Adaptor. *Naturwissenschaften* 68:217–228.

Hartwell, L. H., L. Hood, M. L. Goldberg, A. E. Reynolds, L. M. Silver, and R. C. Veres. 2000. *Genetics: From Genes to Genomes.* Boston: McGraw-Hill.

Margulis, L. 1984. *Early Life.* Boston: Jones and Bartlett.

Maynard Smith, J., and E. Szathmry. 1997. *The Major Transitions in Evolution.* Oxford, England: Oxford University Press.

CHAPTER 6

Behe, M. J. 1996. *Darwin's Black Box.* New York: Touchstone.

Gibson, E. K. Jr., D. S. McKay, K. Thomas-Keprta, and C. S. Romanek. 1997. The case for relic life on Mars. *Scientific American* 277:58–63.

Gonzalez, G., D. Brownlee, and P. D. Ward. 2001. Refuges for life in a hostile universe. *Scientific American* 285:60–67.

Horgan, J. 1996. *The End of Science.* Reading, Mass.: Addison-Wesley.

Kerr, R. A. 2002. Reversals reveal pitfalls in spotting ancient and E.T. life. *Science* 296:1384–1385.

Krauss, L. M., and G. D. Starkman. 1999. The fate of life in the universe. *Scientific American* 281:58–63.

Nicholson, W. L., N. Munakata, G. Horneck, H. J. Melosh, and P. Setlow. 2000. Resistance of *Bacillus* endospores to extreme terrestrial and extraterrestrial environments. *Microbiology and Molecular Biology Reviews* 64:548–572.

Pappalardo, R. T., J. W. Head, and R. Greeley. 1999. The hidden ocean of Europa. *Scientific American* 281:54–63.

Richardson, Lewis F. 1960. *The Statistics of Deadly Quarrels.* Pittsburgh: Boxwood Press.

Index

*What Is Life? The Physical Aspect of the
    Living Cell* (Schröder), 6
Wilberforce, Bishop, 185
Wilkins, Maurice, 64
Woese, Carl, 112

xenon, 57
X rays, 29, 40, 157, 191n3

Yahweh (YHWH), 9, 11
yeast, 7, 72, 77, 80, 137

DATE DUE

GAYLORD

PRINTED IN U.S.A.